New Moon Rising

New Moon Rising

The Making of America's New Space Vision
and the Remaking of NASA

by

Frank Sietzen Jr.
&
Keith L. Cowing

An Apogee Books Publication

THIS BOOK IS DEDICATED TO

Those who Gave Their Lives Exploring the Limitless Frontier of Space

Ad Astra Per Aspera; Semper Exploro

And the Next Generation Who Shall Continue Their Voyage

Audentes Fortuna Juvat

All rights reserved under article two of the Berne Copyright Convention (1971).
We acknowledge the financial support of the Government of Canada through the Book Publishing Industry Development Program for our publishing activities.

Published by Apogee Books an imprint of Collector's Guide Publishing Inc., Box 62034, Burlington, Ontario, Canada, L7R 4K2, http://www.cgpublishing.com

Printed and bound in Canada

New Moon Rising (First Edition) by Frank Sietzen Jr. & Keith L. Cowing

ISBN 1-894959-12-4 - ISSN 1496-6921

©2004 Frank Sietzen Jr & Keith L. Cowing
Photos courtesy NASA except where noted. Front cover photograph taken by the crew of STS-107 and redesigned by Keith L. Cowing.
Back cover photo shows (l to r) President Bush, a White House staffer, Condoleeza Rice, Sean O'Keefe, Paul Pastorek and Andrew Card
Background is 2MASS picture of the Milky Way galaxy showing over 95 million stars.(NASA/JPL)

Contents

FOREWORD

I am from a generation that was raised on the science fiction of the masters. Robert Heinlein, Isaac Asimov, and Jules Verne were just a few of the great writers who kept me up late at night excitedly devouring their tales of adventure in outer space. In 1957, when I was fourteen years old, fiction became fact with the launching of *Sputnik I*, the world's first earth satellite. I was so enamored by this event that I started building my own rockets even though I lived in Coalwood, West Virginia, a town dedicated to mining coal. I was not alone in this endeavor. Kids from all over the country were inspired to build rockets, and plan careers dedicated to the conquering of space. The space race between the Soviet Union and the United States began, culminating in 1969 with victory for the USA when Neil Armstrong placed his boot on the regolith of the moon. The move into space was strong and seemed unstoppable. The Saturn family of booster rockets designed by Dr. Wernher von Braun and his team for the moon-landing program were amazingly reliable, and versatile. We could do anything in space we wanted to do. But then *Apollo* was abandoned, and so were the Saturns. In their stead, an American space truck which promised cheap and easy access to space was proposed and approved. It was officially called the Orbiter, but came to be popularly known as the Space Shuttle.

The Space Shuttles were just starting to fly when I began my work at NASA, and it was an exciting time. The brand-new Shuttle was like no other spacecraft ever flown. Indeed, it would prove to be a machine that could accomplish a wide variety of tasks in space. But as the years passed, we who worked with it began to realize the weaknesses of its design. The Shuttle was always cranky, overly complex, prone to break down, and difficult to prepare for launch. Moreover, the prediction that it would be the cheapest of launchers proved to be erroneous. It was, instead, hideously expensive. Its design flaws would ultimately even prove deadly to its crews.

Long before the Shuttle started to overtly fail, I began to notice frustration among my fellow space workers. No matter how we pretended otherwise, we were disappointed because the Shuttle was an orbital vehicle and lacked the capacity to go to the moon and beyond. Also, with very few exceptions, the experiments the Shuttle carried on board looked down or inward, not up. It was almost as if after the *Apollo* flights, there was an order to NASA to forget the human exploration of space beyond low Earth orbit. The moon and the planets were off-limits, and our astronauts were restricted to carving endless loops around our home planet. A series of presidents did nothing to change this. Clearly, they had little interest in providing leadership in the arena of human spaceflight. We NASA engineers and managers limped along, doing what we could with the Shuttle, and trying to tell ourselves better days were ahead.

After the *Challenger* disaster, and the return to flight that followed, those of us who still worked for NASA gradually lost hope of ever working on a mission as audacious, fun, and challenging as going back to the moon. What we were given, instead, was the International Space Station (ISS). For those of us steeped in space history, ISS and its cobbled-together American-Russian partnership was hardly a dramatic leap forward. Its main goal was clearly an opportunity by the Clinton Administration to support Russian engineers and scientists. This, of course, was for the laudable purpose of keeping them from working on missile systems for governments unfriendly to our country. Even so, we in NASA weren't fooled. ISS was not going to do anything for those of us who wanted to see our astronauts go somewhere farther than a couple of hundred miles off the surface of the planet.

As ISS gradually became the focus of the agency, and the Shuttles became older and ever crankier, NASA morale eroded. After so many redesigns of the space station that we lost count, the program that emerged was a number of habitats and experiment modules bolted together like a tinker-toy, and flying along in a low orbit, with the Shuttle and the 60's-era Russian *Soyuz* used for support. Our government made promises that the Shuttle could not realistically support, while the Russians made promises for the ancient *Soyuz* that their economy could not sustain. It was just more frustration for those of us who longed to work the pathways to space. presidential leadership was desperately needed for NASA, but it seemed it would never come.

When the Shuttle failed once again with the destruction of *Columbia* and the loss of her crew, NASA's human spaceflight program lurched to a halt. A committee was appointed by President George W. Bush to find out why. The result was a focus not so much on the Shuttle but on the culture that had grown up around it. It was a culture, the committee informed congress and the president and, indeed, the American people, that had little accountability, and ignored clues that the Shuttle was a dangerous beast. Essentially, NASA engineers and managers were cast as thick-headed and arrogant. Frustration and angst naturally flowed like a river through the agency.

I was lucky. By then, I had been away from NASA for years, happily sustained by a writing career that included *Rocket Boys/October Sky*, the story of how I'd built my rockets back in Coalwood when spaceflight was new and the possibilities seemed unlimited. The Shuttle had brought those who still believed in that dream to their knees, and I was glad I wasn't part of it any more. Yet, I still cared, and hurt for the space agency troops. Once a part of the NASA's family, always a part, I suppose. Then I heard President Bush was going to make a major announcement on space, one that would set a new course. I had recommended that he do so three years earlier in a Wall Street Journal editorial (see http://www.homerhickam.com), but I also recognized that the war against terrorism had kept the president more than a little busy. Pragmatism is one of my strong suits. Late is always better than never.

The day of the speech finally came. I listened to the president and concluded it was a good speech, and said a lot of what I and NASA engineers and managers had wanted to hear for a long time. In fact, the more I thought about it, I realized President Bush had demonstrated leadership on a scale not seen since John Kennedy. We were going back to the moon, and then on to Mars, and it was clear he meant it. To accomplish these goals, there was also a promise that the Shuttle was going to be replaced (*Finally!* was my thought) by a vehicle, or vehicles, which could take humans anywhere a future president might like them to go. It was a reasonable, practical proposal on how to take the next steps in space. I was proud of the president that day. I still am, though my pride is tempered by the reality of a powerful opposition against him and his vision.

The book that follows details the evolution of the president's proposal, and the men and women who conceived its details, and how they did it. It is an interesting tale, and instructive. It is even inspiring. Still, we of NASA, or once of NASA, harbor hope, but few illusions. The forces against moving out into space are powerful. Many of the nay-sayers sit in high public offices, and guard their public funding satraps with ruthless jealousy. The moaning and groaning over the president's plan has only just begun. It is therefore up to all of us who believe in the potential of human spaceflight to fight for this vision. If we don't, I'm convinced the price we'll pay is stagnation, not only of our human space program, but of our country. To be a great nation, a nation must do great things. Greatness for us, I believe, lies along the pathways to space. The choice is stark, and the stakes couldn't be much higher. America's leadership in the world, and the survival of our values, may depend on how we choose.

— **Homer Hickam,** Huntsville, Alabama, May 29, 2004

AUTHORS' NOTE

This book is the story of a remarkable period of historic change in the U.S. civil space program. Specifically, it brings the reader into the senior management of the National Aeronautics and Space Administration as that leadership faced an unprecedented crisis. That crisis would give rise to an opportunity to elevate the discussion of space policy and space activities to the national stage. That discussion in turn would yield a new, comprehensive space vision for the nation. Whether that vision ever becomes translated into spacecraft and missions is the challenge that now faces NASA.

This new vision and mission for NASA would not come without a price. As the book details, NASA administrator Sean O'Keefe and his leadership team would be forced to literally reinvent the space agency to free up funds to pay for much of the space plan that President George W. Bush would propose. He would also attempt to install a management structure more aligned to innovation and business-like practices than NASA's more traditional defense of its status quo.

Moreover, O'Keefe would have to do this as the agency emerged from a decade where budgets were cut year after year, and management was often more interested in looking for scapegoats than finding new worlds.

The reader is also taken into the counsels of discussion inside the Bush White House, as three groups of staff and senior appointees thrash out the eventual direction for the president to take in space matters.

This is the story of what these people saw, heard, thought and felt as these events unfolded around them. It is not a comprehensive review of these events. It is not a scholarly history of space policy, space politics, or how the Bush administration functions. It is not a detailed history of NASA, or the O'Keefe administration's management of it. The authors make no attempt to judge their actions as being good, bad or indifferent to the nation's interests. That is the task of the reader.

What we have attempted, to the extent possible given the time available, is to insert the reader into the middle of these historic events as they unfolded. We believe, therefore, that we have captured a sense of what it was like to lead the space program from 2001 to the summer of 2004, a time when circumstance, political interests, determination, and history collided. The result was to transform a handful of leaders and bureaucrats into change agents.

That change was about to be forced upon NASA in the skies high above Texas, on February 1, 2003. Still to be decided was just how far this change had progressed by the summer of 2004, and the prospects for additional changes in the future.

This is the story of how it began.

PROLOGUE:

ACROSS A HINGE OF HISTORY

With a glint of sunlight and a pair of sonic booms, the winged ship was to announce its arrival in the skies above the Kennedy Space Center. Now, just before nine a.m. on the morning of the first of February 2003, the senior leadership of the civil space program was arrayed near its runway, awaiting the descent and touchdown.

A crowd of shuttle crew families and NASA officials made up a large part of the group, but one man in particular was key. That man, the one that many wanted to speak to, was a tall Irishman with receding, prematurely white hair. He was slightly stooped sometimes when he walked, but his mustachioed face betrayed a boyish excitement about all he saw, crinkling into a laugh that came in rapid, short bursts.

On this morning, he had every reason to smile. Since taking over the leadership of the National Aeronautics and Space Administration two years before, he had achieved a major turning point for his agency. By a combination of cuts to some of NASA's most cherished programs, and a bit of budgetary slight of hand, he had staved off financial disaster and ended the space agency's chronic annual bouts of cost overruns. His success was noted by the man who appointed him to the task, the 43rd president of the United States. George W. Bush had sent him to NASA with a clear and compelling mandate. The president was blunt. The place, he said, was messed up. Go fix it. He had started doing just that.

And, in the process, dragging his agency out of what some were calling the 'penalty box', creating an opportunity to consider something that NASA hadn't received in decades: possibly a new mission. He had already obtained permission from the White House to develop the first all-new human spacecraft in three decades. Called the Orbital Space Plane, its development would buy back the opportunity to expand the crew size aboard the International Space Station, as well as take some of the burdens off the oversubscribed fleet of Space Shuttles.

His adroit handling of the political and budgetary issues that confronted him during these two years hadn't gone unnoticed on Capitol Hill, where he had worked for years as a Senate aid before heading to the Pentagon for a tour of duty as Comptroller and Chief Financial Officer, and later, as Secretary of the Navy. In that job, he was sent over by then-Defense Secretary Dick Cheney to clean up a sexual harassment scandal. When that administration left office, he returned to a lifelong love, teaching, at Syracuse University in upstate New York, a far cry from the New Orleans of his birth.

And while he worked through restoring NASA's reputation with money and manpower, he brought with him-or promoted from within-a small band of revolutionaries, mainly from outside the space industry, who could help him advance Bush's management change agenda. Several of them were gathered with him that morning, a day that was supposed to be one of triumph and renewal as the Space Shuttle Columbia swooped down from orbit and landed in front of them.

The nation's oldest shuttle, the Columbia, had made its first landing at Kennedy Space Center in March 1979, riding atop its 747 carrier aircraft. Now, in about 16 minutes, it was scheduled to make its latest arrival there, at what was its home port. The smiling Irishman chatted amicably with several NASA and industry workers, including one old and former co-worker. It was a cool and beautiful morning, the kind you remember. He saw one of his most trusted aides, Associate Administrator for Space Flight Bill Readdy approach. Readdy, another Irishman, was a former astronaut who had flown into space aboard the shuttles. He smiled as his colleague approached, until he noticed that Readdy's face had turned ashen.

And then, Sean Charles O'Keefe's life was turned upside down. Space shuttle Columbia and its crew were gone, and the world that he knew went with it.

By the time he returned to the Capital, after 11pm that night, that day of searing loss and tragedy was nearly at an end. The next morning, Sunday, would be a new day. But while it was the beginning of another day on the calendar, in reality, it was an end.

On history's clock it was sunset. An era spawned three decades before—the era of winged reusable spaceflight—had ended abruptly in the skies above central Texas the previous morning, although no one yet sensed or understood it. The destruction of orbiter 102, the Columbia, in front of the whole world, and the deaths of seven more shuttle astronauts would expose a fault line within NASA that would shake it to its very core. It would also expose the shuttle as fragile and aged, although few would say so in public throughout the year to come. By the time the board investigating the Columbia accident had completed its work, and released its report in late August, no shred of the shuttle's illusion as a space-going truck would remain.

That same investigation would castigate NASA's leaders for failing to either replace the ailing shuttle fleet, or come up with a politically acceptable rationale for the existence for the civil space program. Faced with a politically-imposed requirement to strengthen the shuttles and to recertify them, should NASA wish to continue using them beyond another decade, NASA would choose instead to bring their use to a swift end. In their place would be neither a replacement craft of similar capability, or some other form of winged space truck. Instead, driven by the fault lines exposed by the accident, and a bipartisan call for a new space vision, NASA would propose replacing the grounded shuttles with a type of craft that the shuttles had been designed to replace: space capsules.

Those capsule designs would call America to an extended exploration initiative once only dreamed of by space supporters since the end of the Apollo era: send humans beyond Earth orbit again to explore the Solar System—starting with the moon. That call would be made by the most unlikely player of all: an American president who had shown virtually no public interest in space activities, either as a Texas Governor, presidential candidate, or sitting president. More importantly, in choosing extended spaceflight as his goal, George Walker Bush would face hostility and opposition that would need to be overcome. His main instrument to fight that war and win the new space program was the Irishman from New Orleans.

But none of that path lay obvious as O'Keefe and NASA began to sift through the Columbia wreckage, the physical and the political.

By Monday morning he had arranged to have a large group of Senators and Congressmen ready for a NASA briefing. But soon after O'Keefe arrived at his desk, Andy Card, the president's Chief of Staff was calling from the White House. "The President wants a briefing on the latest," Card said. O'Keefe, expecting Bush to come on the phone, waited for a few seconds. But Card spoke up, "He wants to see you."

O'Keefe dialed Stevens, and asked that the congressional meeting be postponed. Stevens said he would see what he could do, and shortly thereafter called back to say he had rescheduled the meeting. But he also told O'Keefe that several members were unhappy that their schedules had been disrupted. "You just say," he told Stevens, "that I guess the President pulled rank on them all..."

All were now on the other side of a hinge of history; a great technological divide whose resolution would shape the direction in which Americans would explore space for a generation. It was not the first time the nation had faced such a choice. It was a time, as writer Theodore H. White had once said about another era, "a moment of history that creases the memory." Bought and paid with blood and treasure, born in the fires of the second shuttle accident was opened a new pathway for the American space program. Driven by a national discussion centered in the Congress, President George W. Bush and his administration would move across 2003 to come to terms with issues long-delayed or ignored when it came to space policy: what was the reason for the U.S. civil space program? What was the Space Shuttle's role in it? And, ultimately, what were the nation and its leaders willing to spend to accomplish whatever goals were approved to pursue?

Some 40 years before, another American president would be confronted with a hinge of space history. Then, the survival of the nation itself was at stake. One foreign policy debacle after another embarrassed the young administration of John F. Kennedy into seeking a peaceful way to compete with Communist expansionism. What emerged then was a space race to the Moon. Kennedy placed a huge amount of his political capital behind what became Project Apollo. His reason had little to do with the value of manned lunar exploration. Its rationale was instead politics and foreign policy.

Now, with Sean O'Keefe at the helm, NASA was struggling to reform itself. But unlike Kennedy, Bush himself seemed to want to seek a purpose for space exploration separate from its foreign policy rationale. That earlier geopolitical purpose had served during the Clinton years to save the space station from threats of cancellation. As the aftershocks of Columbia reverberated through Washington and the country, O'Keefe was drawn into a review over space policy at a depth no one would have thought possible from the indifferent Texan president.

Bush wanted to fix NASA. But could it be fixed? Was the shuttle 'accident' just that, an unavoidable accident, or had this latest shuttle disaster been caused yet again by mismanagement or even negligence? And when the time came to draw the cause and cure into legislation and federal dollars, would the 43rd president-like his long ago predecessor-be willing to expend some of his own political capital and budget money to set space exploration on a renewed course?

The wounds of that dark Saturday in February were still fresh as the space leadership of the country drew itself together. First they had to fix the shuttles, then return them to space. But while they were trying to do that, the administration had to define their role not only in the current space program, but in whatever effort it chose to follow-or perhaps decide instead that there was no longer a need, in the early 21st Century, for a space program at all.

America's space program was expensive, and in some ways, decrepit, and had been amazingly neglected for decades, requiring a fresh infusion of cash just to become competitive again-all steps that this administration studiously avoided taking in other areas of federal activity.

Budget deficits were skyrocketing. One war was being fought and a second was about to take center stage. Medicare needed reform. So did the nation's energy grid, whose shortcomings were about to be spectacularly demonstrated. Everywhere one looked, there were areas of neglected federal need crying out for spending, from homeland security to domestic issues. In such a mix, who really cared about a collapsing space program?

In February, 2003, America stood poised at a fork in the road. Space exploration could be a major path for the United States-or be defined as a quaint remnant of the Cold War that was too expensive or complicated to continue. The outcome for NASA, as the year 2003 began, might well be everything-or nothing. The story of that extended year; from February 2003 to June, 2004 would be a tale of reconsidering old dreams and attempting to forge new alliances to achieve them. As it began, the cast of characters was centered around a new NASA administrator, without a space pedigree, heading a dysfunctional agency that many believed was falling apart. On the morning after the Columbia accident its future looked as bleak as that of the Space Shuttle, and a new consideration of space policy a distant dream.

Of all of the ten men that would be chosen to lead the national space program, Sean O'Keefe seemed the most unlikely agent of institutional change, and Bush as his president was hardly thought of as another Kennedy.

Thus said conventional wisdom.

*   *   *   *   *

1

THE LONG GOLDIN GOOD-BYE

Part of the groundwork that led to the Bush administration's new space policy was a dual-track process. Part one was bringing to an end the longest serving NASA Administrator in U.S. space history. Part two was finding someone to replace him. That Administrator was Daniel S. Goldin.

A former TRW defense systems manager, Goldin had been plucked from the corporate world by President George H.W. Bush in the early spring of 1992 to replace Navy Admiral Richard Truly, whom Bush had fired in a politically embarrassing and public dispute over the future of NASA. At the time, the space agency was in another period of demoralization. A major space initiative attempted by Bush had been stillborn in the Congress after having been abandoned by NASA itself.

Called the 'Space Exploration Initiative' (SEI), the proposal called for returning astronauts to the Moon's surface and sending them 'on a journey into tomorrow' to Mars. But astonishingly, Truly had little interest in selecting the new ambitious goals for NASA, and in fact seemed preoccupied with more mundane issues such as maintaining the shuttle fleet in service and getting construction of the long-delayed space station underway.

A 90-day study of Bush's proposal ordered up by Truly attached massive development programs for advanced spacecraft and heavy-lift launch vehicles to Bush's goals. The cost of the program, believed to be in the hundreds of billions, was widely derided in Congress and was a prime reason why the project was eventually abandoned. Indeed, SEI was dead on arrival when it reached Congress.

Enter Dan Goldin

In the aftermath of the SEI fiasco Bush turned to Dan Goldin to restore management stability and direction to the civil space program. The new Administrator went about doing just that, in a shake-up management style that sought to re-orient NASA to the political and funding realities of the shuttle/station era—not the long-ago time of Apollo. Budgets were unlikely to grow much beyond current spending levels, despite a flurry of recent reports calling for substantial increases to

NASA's budget and the establishment of new, far-reaching space goals. Goldin was a pragmatist, brought in by Bush and Vice President Dan Quayle to get NASA under control. Alas, Goldin wasn't at the top of their list.

Throughout his decade-long tenure, Goldin imposed a series of management reforms upon his agency. He canceled its pattern of developing large and expensive single-purpose science probes and missions, and shocked the space industry by demanding more accountability from its contractors. "Stop crying to your Uncle Sugar," he told one industry luncheon in 1993.

As President-elect Bill Clinton prepared to take office, little had been said about space activities by him during the 1992 campaign. Former administrator Truly had endorsed him, along with other former Bush officials, such as retired admiral William Crowe former chairman of the Joint Chiefs. But while Truly denied that his public embrace of Clinton's candidacy wasn't a political payback for Bush firing him, he did not succeed in gaining any interest in space from candidate Clinton.

Normally, the departing appointees of the outgoing administration use the transition period to complete unfinished business, make the round of good-bye parties, and otherwise prepare their offices for their successors.

Not so with Goldin.

Instead, Goldin managed to survive the Bush-Clinton transition by adroit political skill that took advantage of the incoming administration's lack of interest in space. Goldin also seized upon a space station descoping and redesign that many in the new administration hoped would lead to the project's cancellation.

Instead, Goldin managed the entry of the nascent Russian Federation into the project and created a geopolitical partnership that the new Clinton-Gore administration came to support. The result was Goldin became a rarity in Washington—a Bush appointee that remained in the Clinton administration. The fact that Goldin had used smoke and mirrors to portray Space Station Freedom in a less than honest light—as well as bare-knuckle internal politics to force his workforce into compliance, in order to push through the new space station design—would come back to haunt him a decade hence.

His methods not withstanding, Goldin's reforms would continue into the latter part of Clinton's second term. But by 1996, Goldin's abrasive style began to rack up increasing political critics—internal and external to the agency. The troubled Russian space station partnership became a flashpoint in Congress. Goldin's response to the failure of the Clinton administration to increase the agency's annual budgets was to increasingly cannibalize other NASA programs to pay for cost overruns on the station.

Indeed, having cheered declining budgets for the better part of a decade, Goldin could hardly do an about-face and say that he needed more money—so he sought it elsewhere. A main target for these cuts would be the space shuttle as well as some NASA science programs, the degree and depth of these cuts only becoming clear in the years after Goldin's departure.

By the time of the 2000 election, Bush campaign officials had made it clear if they were successful, Goldin would be replaced. But with the results of the election unclear, Dan Goldin clung to his position using every skill he had developed over the years. Right now, with the chaos swirling about the 2000 election the best tactic was the simplest: keep a low profile—just as he had done during the course of the election.

When it became clear that George Bush would be declared the winner of the election, NASA's senior political appointees did what is normally done when an Administration is about to end or change parties: you leave. In the process, you appoint someone to act in your stead, usually a career civil-servant.

Among those orchestrating their own prompt, and professionally executed departure were Peggy Wilhide, Associate Administrator for Public Affairs, who came to NASA from Vice President Gore's office. After taking some time off, she would go on to become Vice President of Communications for the Association of American Railroads in June 2001. Lori Garver, Associate Administrator for Policy and Plans, also active in Democratic politics would eventually join the Washington Lobby firm DFI International and briefly compete against boy-band singer Lance Bass for a commercial ride on a Russian Soyuz to the space station as 'AstroMom'.

Roberta Gross, NASA's Inspector General, and Arnold Holz NASA's Chief Financial Officer also announced their resignations.

But not Goldin.

Clinging to power

Having been appointed by George H.W. Bush and then kept on by Bill Clinton, Goldin was not going to leave the office without being shoved. The phrase "Administrator for life" was often heard outside his presence when describing how he viewed his job.

As inauguration day approached, many within the agency wondered why Goldin had not submitted a letter of resignation and designated an interim (acting) Administrator. Had he already cut a deal with the incoming Bush Administration? Or was he simply convinced that they'd keep him on? Or was he just clueless about what was going on? No one was quite certain—except that Dan Goldin was going to sit there until someone said, "We don't want you," to his face.

Apparently. It worked.

On 22 January 2001, two days after the inauguration, Goldin announced the names of career civil servants who'd serve in an acting capacity until the White House found replacements: Steve Varholy would be Acting Chief Financial Officer; Mary Dee Kerwin would be Acting Associate Administrator for Legislative Affairs; Paula Cleggett would be Acting Associate Administrator for Public Affairs; and Beth McCormick would be Acting Associate Administrator for Policy and Plans. Courtney Stadd would begin to serve immediately as NASA's Chief of Staff and White House Liaison.

A press statement issued by NASA HQ announced, "[Goldin] has agreed to stay on until the new Administration makes some final decisions about the direction of the space agency." Goldin was quoted as saying, "I serve at the pleasure of the president of the United States. I am pleased to be of assistance during this transition process. I am committed to making this transition smooth, and I expect that every employee will support that goal."

Will someone please say YES?

Searching for a replacement for NASA's Administrator was not at the top of the incoming Administration's list of priorities. Indeed, given that the whole transition process had been in a state of suspended animation during a large part of the November - January period wherein such searches are undertaken, even less time was available to focus on seeking out Goldin's successor.

Speculation began to mount about whether Goldin clearly understood that political appointees traditionally submitted their resignation at times like this. Some wondered whether the new White House would have to order Goldin out the door. Eventually it became clear that Goldin was not going to be gone on Inauguration day: he had submitted a letter to the incoming Administration offering to serve and naming himself in the capacity normally reserved for a subordinate during the transition.

The Bush White House was preoccupied and saw this as a headache they did not need to bring upon themselves and took Goldin up on his offer.

But they were not going to let him have free reign.

One of the names that had circulated as a possible replacement for Goldin was Courtney Stadd. Stadd was well-liked, had served in a variety of space-related positions inside and outside of government, in the private sector, and elsewhere— and was seen by some as a desirable choice. Stadd was somewhat less than enthusiastic about the prospect of being NASA Administrator having just started a new company outside the beltway.

But he was convinced to come to the agency on a short-term assignment as White House liaison and chief of staff. Stadd immediately began his tenure by telling people that this was a job that would last only a few months— until the Administration was up to speed and a permanent successor to Goldin could be found.

In reality, Stadd was brought into NASA to keep Goldin under control.

Despite his avowed preference for a short-term stay at NASA, Stadd would eventually stay on well past Goldin until 2003. He was soon joined by long time friend Scott Pace who came over from the RAND Corporation.

Nominée du jour

As the new administration continued to fill open positions, the job of NASA Administrator seemed to get scant attention. One name had circulated with some frequency in late 2001: former New Mexico Senator and Astronaut Harrison 'Jack' Schmitt. It wasn't so much that Schmitt was seen as an obvious answer as much as it was the fact that his name had been tossed into the round-robin that the media tends to stir up at times such as this.

In this particular case it happened in a wire story just before Christmas. With everyone heading out of town, no one was around to knock Schmitt's name off the list. As such the story continued to circulate-Schmitt's name stayed in play. People started to get used to hearing his name. Schmitt would later say that he had received several calls during that period-but that they were from mid- to low-level staffers who were seeking to confirm his location and how to contact him.

A rather humorous consequence of this way of circulating names occurred in January 2001. The National Air and Space Museum had hosted a special event honoring science-fiction author Sir Arthur C. Clarke and the film '2001: A Space Odyssey'. Part of the festivities included a showing of the newly restored film at the old location of the Newseum in Rosslyn, Virginia.

Opening the showing was a taped presentation made several weeks earlier by Clarke at his home in Sri Lanka. Not knowing precisely who would be in attendance, Clarke had spoken to who he though might be there. Clarke welcomed, "Dan Goldin," adding, "I understand your successor Jack Schmitt is there with you." This resulted in a large chuckle in the audience given that Goldin was not present—and Schmitt was—and he was as much in the dark about the process as anyone else.

The explanation for Clarke's comments may be what Space Frontier Foundation's Rick Tumlinson has suggested. Tumlinson had visited Clarke in Sri Lanka several weeks before and had given Clarke a lay of the land with respect to the political landscape. At the time Tumlinson was traveling, Schmitt's name was still echoing in the media.

But soon Schmitt's name dropped off the media's radar screen. Henceforth, various names would pop up and then fade away over the coming year. Several of the names floated were by the individuals themselves, to no avail. It would seem that no one wanted the job. And those who wanted the job were not what the White House was looking for.

While no one was angling to get rid of Goldin—no one was arguing to keep him on either. As such, Goldin had to run the agency on a day-to-day basis. No big new projects. Gone were the grand, confident statements. Instead, Goldin spoke far less frequently in public, and when he did it was in measured words always supportive of the White House. Goldin's credibility, mostly in terms of his tenure at the agency which was now on its way to being the longest of any administrator, was coming increasingly into question.

Goldin's tenure unravels

Shortly after the November 2000 election, word began to filter out of NASA that a massive, multi-billion dollar cost overrun had been uncovered in the space station program. When it was finally admitted, many asked why Goldin had suddenly learned of it now—months after the problem began—and, more cynically, why news emerged only after the election. Had Goldin suppressed the news so as not to hurt Gore's chances? Were his own people keeping him out of the loop?

In early 2001, during a break in a NASA Advisory Committee meeting, Sean O'Keefe, still Deputy OMB Director told reporters in a hallway that Goldin "had to know" that the cost overrun was happening.

Goldin also saw some of the programs he had once championed as indicative of NASA's new way of doing business on the chopping block.

On 1 March 2001 it was announced that the critically wounded X-33 program and the less ailing, but still hurting X-34 program would receive no further funding by the Space Launch Initiative (SLI).

In the case of the X-33 program, Lockheed-Martin, NASA's industrial partner had the option to continue the program with its own money. However, given that the company had been forced to abandon the notion of a wholly commercial follow-on to the X-33 (Venture Star), stating that this could only happen with government funding, there was little enthusiasm to spend one more dollar on the X-33. As such the program was effectively dead.

The X-33 program had been announced with great flourish in 1996 by Goldin and Vice President Al Gore at NASA Jet Propulsion Laboratory.

Now the biggest issue was who would get the scraps.

The 800 pound gorilla

Goldin was finally free to deal with a personnel issue that had vexed him for some time. On 23 February 2001 NASA announced that Goldin was appointing NASA Johnson Space Center Director George Abbey to be his 'Senior Assistant for International Issues'. Roy Estess, Director of NASA's Stennis Space Center was appointed to serve as the Acting Center Director of JSC.

Sources inside NASA Headquarters did not suggest anything more complicated about the reasons why Goldin axed Abbey, except to say that Goldin did it "because he finally could." Suggestions that the White House had pushed Goldin to oust Abbey were not in evidence.

Outside the agency, Abbey's 'reassignment' was seen for what it was: he was fired. Given the way Abbey had grabbed control of the Space Station program and dragged it back to JSC, many saw this move by Goldin as the first step in dealing with years of cost overruns and schedule delays.

Abbey would find himself appointed to several advisory positions and eventually moved to palatial offices in the Boeing building just outside the gate at JSC. While gone from line management at NASA, Abbey would continue to cast a long shadow that would still affect the actions of agency personnel for years to come.

Abbey would surface periodically. First, he popped up as a senior fellow at the Baker Institute and served as NASA's representative to the World Space Congress which was held in Houston in October 2002. After a year of remaining mostly invisible he would formally announce his retirement on 7 November 2002 when it became very clear that NASA no longer desired his services—in any capacity.

Commenting on Abbey's departure, Sean O'Keefe noted, "George is a demanding leader who rarely accepts compromise. His ability to motivate and inspire his staff to work harder and smarter helped NASA write much of its human space flight history. His devotion to the success of America's space program is unquestionable and I wish him the best."

In January 2003 Abbey was named Chairman and Chief Executive Officer of American PureTex Water Corporation and PureTex Water Works.

Still no luck

One would think that being at the helm of America's space agency would be a job that many who grew up in the space industry would savor.

But such would not be the case. Some of the people approached for consideration would reply incredulously along lines such as, "You want me to give up several hundred thousand dollars a year, move to Washington, and get yelled at every day as I clean up after Goldin? No way."

As 2001 wore on, the low grade search for Goldin's successor continued. At one point Pete Aldridge was under consideration. Already angling for a position at the Department of Defense, Aldridge was not interested.

One of the more unusual names to emerge was that of Oracle CEO Larry Ellison. Known for his strong personality and management style, Ellison expressed interest in the job. Getting a prominent CEO to join the Bush Administration would be in keeping with previous appointments, most notably Treasury Secretary Paul O'Neil, former CEO of Alcoa. Ellison would eventually offer that he'd take the job so long as he could spend a certain amount of time on the West Coast. This was a non-starter for the Bush White House. Whoever was going to take over NASA would need to be at NASA Headquarters-where the problems were 24-7-not in Silicon Valley.

One last time

Sensing that the days ahead were numbered, Goldin began a victory lap of the agency. One aspect of this last chance to grab a piece of something was the showering of awards. On 30 March 2001 the National Space Club presented Goldin with the Dr. Robert H. Goddard Memorial Trophy. Past winners included John Glenn, Wernher Von Braun, and President Ronald Reagan.

On 5 April 2001 Goldin began his 10th year as NASA Administrator. A few weeks earlier Goldin had passed the previous record for time on the job as NASA's leader. A crowing press release issued by NASA Headquarters noted, "Within the past two weeks, Mr. Goldin's legacy of outstanding leadership has been honored with three prestigious awards. Recognized as one of the visionary leaders in minority higher education, the Administrator received the Federal Leadership Award from the National Association for Equal Opportunity in Higher Education. The Metropolitan Washington Chapter of the ARCS Foundation, an organization dedicated to filling the Nation's need for scientists and engineers, honored Mr. Goldin with the 2001 Eagle Award."

The next day the Metropolitan Washington Chapter of the Achievement Rewards for College Scientists Foundation presented Goldin with the 2001 Eagle Award.

The curious case of Jerry Brown

On 10 April 2001 one of the oddest events in Goldin's tenure began. Jerry Brown (not the California politician), described in a press release as "a senior corporate communications executive" was named the Associate Administrator for Public Affairs. According to that release, "In 1992, Brown was appointed deputy director of the Office of External Affairs for the U.S. Agency for International Development for the Administration of President George Bush. In that capacity he managed press relations. Brown also served as the director of public affairs at the Federal Transit Administration for the U.S. Department of Transportation."

Brown clearly had the credentials, and the political background, to be given the job.

Then suddenly on 27 July 2001, Brown was gone. The 7 August 2001 issue of PR Week said, "NASA head of public affairs Jerry Brown abruptly resigned his post on July 27, the latest casualty of what a number of former NASA employees are calling intense pressure from senior staff to award outsourced projects in an anti-competitive manner. According to sources speaking on condition of anonymity, Brown was one of many employees at the agency encouraged to feed new projects to companies with existing NASA contracts rather than let them be offered on the open market. Two other high-level employees who have recently resigned cited this 'unethical' situation as the reason behind their departures."

Conflicting stories circulated at NASA Headquarters regarding Brown's departure. Some say he quit, others suggest he was asked to leave. Still others suggest that Brown had come to realize that he had walked into a nuthouse (NASA HQ) and that it was time to escape. Regardless of the reason(s), Brown's departure only served to underscore the growing perception that NASA was an agency with systemic problems and that it was adrift and in desperate need of some leadership.

With Goldin in the eighth month of his caretaker status, that leadership was nowhere in sight. From Goldin's perspective, the less controversial he was, the greater the chance he'd be considered to continue to serve as NASA Administrator.

Shortly after Brown departed Goldin appointed his personal PR man, Glenn Mahone, to be acting head of NASA public Affairs. He would then appoint Jeff Bingham as the Acting Associate Administrator for Legislative Affairs. Bingham had recently provided support to the Bush-Cheney Transition Team and had been previously appointed to be a Special Assistant to Courtney Stadd in January.

Dealing with the unavoidable

By July 2001 Goldin could not hold off external criticism of space station budget problems any longer. On July 31st he appointed the ISS Management and Cost Evaluation Task Force (IMCE) which would also be come known as the Young Commission after its chair, frequent external investigation panel chairman, Tom Young.

According to the press release announcing its creation, the IMCE was chartered to "help NASA address the recent cost growth on the program by assessing the quality of the ISS cost estimates as well as program assumptions and requirements, and identifying high-risk budget areas and potential risk mitigation strategies."

In reality Young's panel would end up painting a damning portrait of a program long neglected and in serious need of dramatic overhaul. The IMCE's recommendations would continue to echo through the agency for years to come—well after the Columbia accident in 2003 and the announcement of George W. Bush's space policy in 2004.

That last day of Innocence

On 6 September 2001, on one of those stunningly crisp, cool days Washingtonians crave, a hearing was held at the Russell Senate Office building on shuttle safety.

At the time NASA felt that it was on a secure path to upgrading the shuttle fleet such that, in the words of Bill Readdy, then Deputy Associate Administrator for the Office of Space Flight, who testified at the hearing "it will enable the Space Shuttle to fly safely well into the second decade of the 21st century." Meanwhile, Richard Blomberg, chair of the Aerospace Safety Advisor Panel warned, "The long-term situation has deteriorated." Yet he also felt, "NASA needs a reliable human-rated space vehicle to reap the full benefits of the International Space Station (ISS), and the Panel believes that, with adequate planning and investment, that vehicle can be the Space Shuttle."

The Senators present were concerned about the situation, and pressed Readdy as to why certain upgrades were being deferred or why more funds had not been focused on dealing with shuttle upgrade and safety issues. But no one was waving their arms about suggesting that any imminent threat existed.

Upon leaving the hearing Readdy, noticing the stunningly blue cool sky, decided to walk back to his office. That walk ended up becoming an hour long chat with a reporter on the status of the space program.

At that time Readdy saw that the ISS, while somewhat strained to meet the interest of a wide variety of users, could easily be tweaked such that it served as a test bed for the flight qualification of humans for long duration voyages. The tweaks would involve picking the research. No significant costs were envisioned.

On that day Readdy was optimistic about where things could go, but like everyone else in the agency, was tired of what NASA had been put through over the past decade. Cost overruns on the ISS certainly did not help matters.

As he excused himself and headed back to his office for a meeting, little did Readdy, or anyone else, know what would happen 5 days later on a cool, sunny September day. Readdy's home was a few blocks away from the Pentagon.

Some 250 miles above the Earth, the International Space Station sped along as September 11th arrived. Onboard, Expedition Three Commander Frank Culbertson, Pilot Vladimir Dezhurov and Flight Engineer Mikhail Tyurin were in the fifth week of a four-month stay aboard the station.

Preparations were under-way for the arrival of the Russian PIRS docking compartment due for launch in 3 days and arrival at ISS several days later. Otherwise, all things had been moving along at a normal pace.

In Washington, and all along the east coast, it was another one of those classic clear September days Readdy had stopped to admire the previous week. While many were still on the commute into work, the attacks began. At NASA Headquarters, as with the rest of Washington DC, no one quite knew what was going on. Soon people were watching news footage of the Twin towers on fire in New York. Then came word of an explosion at the Pentagon and rumors (which later proved to be false) of another one at the State Department.

Anyone looking to the west from Washington could clearly see a plume of dark smoke rising from across the Potomac. Meanwhile, rumors of another plane flying up the Potomac toward Washington made the rounds. Other rumors spread of an odd plane seen circling above the Mall.

People quickly left their desks and, in the hours ahead, managed to find their way home. Soon the entire agency would either be shut down or shut off from the rest of the world. Shuttles were secured, and gates were locked.

Bad news from home

The news from Earth that morning wasn't good. Frank Culbertson would soon find that some of the day's pre-planned routine would be altered. As soon as he was told of the attacks, Culbertson checked to see when they would be passing over the east coast of the U.S. Discovering that this was only some minutes away, Culbertson grabbed a camera. The window in Mikhail Tyurin's cabin turned out to be the one with the best view.

Over the next several days Culbertson would shoot still and video imagery out of the station's windows as it passed over New York City. Space Imaging would follow with high resolution imagery of the damage in New York and Virginia taken by its IKONOS satellite. Both the close-up and distant views from space were haunting.

Culbertson would later radio back to Earth, "Our prayers and thoughts go out to all the people there, and everywhere else. Here I am looking up and down the East Coast to see if I can see anything else, and to the people in Washington."

Later, Culbertson would learn of his own tragic connection to 9-11. The captain of the jet which was crashed into the Pentagon was Charles 'Chic' Burlingame, a former Navy pilot and a classmate of Culbertson's. Culbertson would later reflect back and note how isolated he felt to be the only American off the planet .

Time to go

Barely a month later, on 17 October 2001, Goldin surprised many by announcing that he would be retiring effective November 17th. The announcement came along with a press release which lauded his accomplishments and word that he had "accepted an interim position as a Senior Fellow for the Council on Competitiveness in Washington, as he transitions into the private sector."

The day before Goldin announced his departure, Joe Rothenberg, the popular Associate Administrator for the Office of Spaceflight announced that he would be retiring from NASA effective 15 December 2001. When asked to comment on the simultaneous appearance of these two departures, Goldin was adamant that there was no connection or that they had anything to do with NASA's looming $4.5 billion space station overrun.

Joe Rothenberg had many difficult tasks to confront. The dilemma of the International Space Station being the most constant and frustrating.

However, the most difficult task Rothenberg faced was not technical. It was a task he succeeded brilliantly at: retaining a steady reputation for honesty. This was especially noted on Capitol Hill. Given the aura of distrust that had often surrounded the ever-changing story NASA has peddled on the Hill for a decade under Dan Goldin's chaotic management, this was remarkable to many.

Rothenberg would be replaced by Associate Administrator for the Office of Safety and Mission Assurance, Fred Gregory, who would later find himself nominated as the first Deputy Administrator in more than a decade. Bill Readdy would later move in to replace Rothenberg. Meanwhile, the White House had not publicly anointed anyone as Goldin's successor. With Goldin's departure now a certainty, the rumor-mongering went into high gear.

General Thomas Moorman's name surfaced at one point in Fall 2001 and was on everyone's lips for several weeks. Moorman immediately sought to knock the rumor down. Indeed he suggested that someone was floating his name so as to make it look like he was in a self-promotion gambit thus serving to torpedo a job he wanted at the Pentagon. Former Senator Jake Garn's name also surfaced but he seemed to be content with his retirement and that put an end to the possibility of further consideration.

Now that Goldin was on a clear timeline to depart the agency, the pace of his victory lap picked up. Perhaps one of the more bizarre honors he was bestowed was the French Legion of Honor Award, one of France's highest civilian honors, bestowed on 31 October 2001. NASA described the award as 'recognizes outstanding contributions to humankind' and made certain that images were released of a glowing Goldin having the ornate medal pinned on his jacket by French Ambassador Bujon de l'Estang at a ceremony held at the French Ambassador's residence

Of the event and the award, Goldin was quoted as saying, "It is a rare and distinct privilege to be included among the many illustrious individuals throughout history who have been granted this prestigious award, and I am very honored."

At NASA people just shook their heads knowing he would soon be gone.

An unlikely name emerges

A few days later the rumor-mill suddenly saw a swift change in direction. At a break during a hearing in the Rayburn House Office building, Deputy OMB Director Sean O'Keefe, who had been testifying, was approached about a rumor that O'Keefe had been selected as NASA's new Administrator. "I cannot comment on that matter." was all O'Keefe would (predictably) say.

A week of swirling rumors would follow as word of O'Keefe's selection began to appear. During that period Goldin continued his victory lap at frenetic pace. A large ostentatious exhibit measuring more than 20 feet in length appeared in the main lobby of NASA headquarters. Titled 'Ten Years of Accomplishment' the last decade of the agency's progress had been transformed as a backdrop against which Dan Goldin's accomplishments were highlighted.

Meanwhile, behind locked doors in the Administrator's suite. Goldin's secretarial staff was busy developing an electronic rolodex of all of Goldin's contacts. This database would later be used to allow Goldin to send hundreds of people a small flag which had flown in space and a thank you letter.

Virtually anyone who was anyone in the space community got one of these little gifts from Goldin. Even perennial NASA and space station critic Bob Park from the American Physical Society got one. Park, whose 'What's New?' is required reading among the science policy crowd wrote, "6. NASA: MAYBE DAN DOESN'T READ WHAT'S NEW? Dan Goldin just sent me an American flag that's been on the ISS, to thank me for all my help. Maybe my writing isn't clear enough."

There was also a comprehensive sweep underway to snatch up any memento (not already collected) at NASA that Goldin could take with him. Given the number of awards he had garnered on behalf of the agency (which was often ignored) medals, plaques, and statues were grabbed from walls and credenzas and packed away.

In one instance, a senior Headquarters employee commented how he purposefully hid an award he had accepted on Goldin's behalf when he heard about the treasure sweep Goldin had ordered. The treasure hunting secretaries never found it.

Goldin also sought to take all of his records with him. Here he ran into a road block. His papers were considered official government property and he was not allowed to take them off site, as he had requested with the intention of getting them copied.

Later Goldin was told that he could file FOIA (Freedom of Information Act) requests if he'd like to see these records. Imperial Administrator that he was this did not sit well with the departing space chief.

On 14 November 2001 the speculation came to a halt. The White House announced that it would send a nomination to the Senate asking that Sean O'Keefe be confirmed as the next Administrator of NASA.

Don't let the door hit you

Two days after word came of Goldin's replacement, an elaborate multi-center TV production was aired wherein Goldin bid farewell to the agency, After a short introduction, a standing-room only audience at NASA headquarters (and employees across the nation via television) watched a fancy video presentation highlighting the events that shaped 'the Goldin Years'. To add to the aura of the presentation, none other than venerable space icon Walter Cronkite served as narrator. Every facet of what NASA does got some air time.

When the presentation was completed one was left with the impression that it was Dan Goldin—and Dan Goldin alone who made all of the wonderful things happen. Indeed, with no mention whatsoever of mission failures or cost overruns, you'd think his tenure was smooth and without problems.

Referring to Goldin's tenure at NASA, Associate Deputy Administrator Dan Mulville said, "His hand at the helm has been steady. There has been no greater champion. No greater cheerleader at NASA." Noting that it is a time honored military tradition for a retiring commander to be presented with the unit's colors, Mulville said, "I present you with the NASA Administrator's flag that has been with you since 1992." A few minutes later Mulville awarded Goldin the NASA Distinguished Service Medal—an honor normally bestowed by the Administrator upon someone else.

NASA Public Affairs Chief Glenn Mahone then introduced a series of video clips, taken from around the agency, of people expressing their thoughts about Mr. Goldin. What followed was actually a series of short clips taken only at NASA Headquarters of all of his Associate Administrators and Senior Advisors. Most of the comments were general wishes for a happy retirement and for more time with his family. Others were more detailed-some were rather humorous.

Courtney Stadd referred to Goldin as, "A mentor and a teacher and a great role model as administrator for this agency." Stadd went on to say, "I think we are a better country because of Dan's leadership. Sue Garman then appeared holding (barely) a huge pile of paper action items. NASA Inspector General Roberta Gross gave Goldin an official OIG Fraud Hotline poster. Office of Biological and Physical Research Associate Administrator Kathie Olsen said, "She looked forward to having

Goldin chair a task force at NASA."

After some more comments from the podium, additional taped remarks followed; this time from the NASA field centers. For the most part only the center directors spoke. Some were generic in their comments. Others were downright hilarious.

Johnson Space Center Director Roy Estess appeared in full cowboy regalia in front of former center director George Abbey's herd of longhorn cattle. Kennedy Space Center employees pretended to launch a large hobby rocket with Goldin hanging on for dear life as it streaked into the sky and circled the Earth. Marshall Space Flight Center Director Art Stephenson and some of his employees sang (badly) a 'Cheaper - Better – Faster' version of 'I did it my way'.

Langley Research Center Director Jeremiah Creedon spoke of eliminating the infamous NASA 'worm logo'; a pet obsession of Mr. Goldin's. Creedon said that he and his staff were, "Redoubling our efforts. There is one last worm at the center and we have found it. Sadly, it is in a wall decoration on the wall in my own office." Then as Creedon continued several people came in and removed the offending worm logo; and every other award in Creedon's office (which had been connected together as a stage prop) in one swift action. Creedon beamed proudly that Langley was now "a totally worm-free center."

Last up was Goddard Space Flight Center Director Al Diaz. Diaz started out hopelessly deadpan and official-looking as he sat at his desk and rambled off all of the great things that had been done under Goldin's tenure. After a few minutes he stood up and suggested that his employees weren't always so serious. As he did, outside the window behind him a gigantic NASA Worm logo sign was being hoisted up by a crane as if being replaced atop the Center's Headquarters building. Diaz quickly closed the curtains and looked suitably embarrassed. This was followed by a giant animated worm logo spinning high over the Goddard campus.

Goldin was noticeably twitchy throughout these presentations. Once his center directors had gotten their last jabs in it was time for Goldin to put the spin on his tenure at the agency.

Parting words

Goldin then spoke. He quoted the philosopher Cervantes who said, "He who loses wealth loses much. He who loses a friend loses more. But he who loses courage loses all." Goldin said that this is something that everyone who works on America's space program needs to think about. "You are the privileged few who get to fight in the arena of scientific combat and scientific truth. You need to believe in yourself. Only you know what your capacity is."

Goldin then took on the media and those who appear in it saying, "They do not know the sweetness of success or the bitterness of failure." He advised NASA employees to be wary of the media. "You've got to believe in yourselves, and only you know what your own capacity is—not the critics who sit in the stands and do nothing except look at what you're doing and write in newspapers, speak in the media or testify before the Congress."

Goldin's advice, "Don't succumb to criticism if you believe you're right. That's what this agency is about. Don't listen to the stuff in the newspapers or on the TV. They're main approach is to sell papers and TV time."

"Freedom of the press is essential," he said. "The scrutiny of the press is essential. But you don't have to listen to them if you don't believe they're right. You don't have to respond to it. The way you respond is by overcoming the difficulties." Goldin said that one way people might respond to such criticism is to only set mediocre goals "so that you will always have success and not be criticized." That would be a mistake Goldin said, "Be bold and don't fear failure. Mediocre goals are poison."

Speaking about his successor Goldin said, "On Saturday night I am going to hand the lease of NASA back to the president. I have known Sean O'Keefe for a decade. He will be the Administrator of NASA and he will need your support to make sure that the agency is important. It is not important who leaves the agency: the person who leads cannot do so without your support." He added, "Don't listen to the words of the critics. Listen to the words of Sean O'Keefe."

Goldin then closed by reading from the poem 'The Road Not Taken' by Robert Frost. The poem ends with the line, "I took the one less traveled by and that made all the difference." Goldin added, "Thank you very much for a great run."

Cashing in

It was announced shortly thereafter that Goldin had speaking representation with 'Authorities Speakers Bureau' and was now being promoted as, "NASA's longest-serving Administrator, has joined Leading Authorities Speakers Bureau as a new exclusive speaker. Goldin, who served under three US presidents, is widely credited with reshaping NASA. He is an expert on new technologies, management, Interplanetary exploration and research and development tactics. The National Journal named him, "One of the 100 Most Influential Men and Women in Government," and Defense Business, named Goldin, "Among the world's top 40 most influential defense-industry leaders."

Not losing a beat the press release spoke of Goldin appearances on CBS Evening News with Dan Rather and on the Jim Lehrer News Hour.

Meanwhile the de-Goldinization of NASA had begun: the large photomural of his accomplishments in the NASA HQ lobby was gone—as was his official portrait. Goldin's welcome and lengthy and self-serving biography no longer graced the main page at nasa.gov (although he was still listed as Administrator at www.hq.nasa.gov). Yet Goldin sightings were made at HQ as he continued to clean out his desk.

The grand send off

Of course, this was all leading up to the grand finale, the farewell banquet. On 26 November 2001 NASA alerted its employees of the impending festivities to be held on 4 December 2001 at the Swissotel at The Watergate. Price of admission: $75.00.

More than 250 people attended Goldin's farewell fete. The settings were lavish, standard fare for the soirées that movers and shakers frequent in Washington. The crowd was a curious mix of people from Washington's political, industry, government and media circles.

Among the luminaries in attendance: humorist and host Mark Russell, political talk show host John McLaughlin, the Ambassador from Israel, Rep. Dana Rohrabacher, Rep. Jerry Lewis, Sen. Ted Stevens, Sen. Conrad Burns, former Congressmen Louis Stokes and Robert Walker; and former Senator John Glenn.

From NASA, Center Director Harry McDonald, most of Goldin's Associate Administrators (past and present), ISS Task Force Chair Tom Young, and a number of NASA public affairs and legislative affairs personnel were in attendance. The Canadian Space Agency sent its recently-retired president Mac Evans.

Curiously, very few (if any) rank and file NASA personnel or contractors were in attendance. Indeed there was some concern in the days leading up to the event that there would not be a sufficiently firm showing and some bushes were beaten so as to increase attendance.

Mark Russell served as emcee for the event. While no one else seemed to see the need to adjust the volume of the PA system, Russell felt the need to shout every word. His jokes (even though they were bellowed out in megaphone style) were funny including his initial (mis)introduction as being of "Dan Glickman, the greatest Secretary of Agriculture ..." Throughout the evening, his ad libs were interspersed with glowing pre-written comments that were the obvious product of someone at the NASA Public Affairs Office.

Before anyone else spoke, there was the obligatory showing of the slick video produced by NASA PAO regarding Dan Goldin's accomplishments. This was the same video shown at Goldin's farewell at NASA Headquarters a few weeks earlier.

A parade of politicians then spoke from the podium. All lauded Goldin's accomplishments—saying, in one way or another, that if it weren't for Goldin's determination and leadership, NASA would not be where it is today. They also added remembrances of special times together with Goldin.

Sen. Conrad Burns spoke of Goldin visiting Montana and flying out a plane-load of cherries. Sen. Ted Stevens spoke about multiple fishing trips in Alaska wherein Goldin was by far the most prolific fisherman and where he hogged the shower on a fishing boat while others huddled naked waiting for their turn. Rep. Dana Rohrabacher spoke of Goldin's attempts at surfing in California noting that a special slow motion camera would be needed to present Goldin's entire time on top of the board.

Unlike the one-sided platitudes offered by members of Congress, ARC Center Director Harry McDonald provided a much more accurate and balanced portrayal of Goldin. McDonald, who had once been quietly fired and then rehired in the same day by Goldin over a misunderstanding, spoke also of Goldin's accomplishments at NASA. He went on to recall Goldin's characterization of the way NASA works saying, "Some days you're the pigeon, some times you're the statue."

In describing Goldin's approach to managing NASA, McDonald told a story about famed violinist Itzak Pearlman. At the beginning of a performance, one of his strings broke, but he went on to make an impassioned performance nonetheless. When asked later how he managed to do this, Pearlman said, "Sometimes the artist is required to make beautiful music with what he has at hand."

McDonald then presented Goldin with an award made by one of Ames' wind tunnel technicians "in his 'spare' time - something he has a lot of these days" McDonald said, noting Goldin's insistence that Ames move away from wind tunnel operations.

Astronaut Steve Lindsey, commander of STS-104 was next. He spoke of the marvels he saw on his recent mission to the International Space Station, and noted that this would not have been possible without the administrator's devoted efforts. He went on to describe the telecons Goldin has with crews prior to their launch. Being a rookie the first time he experienced this, he thought Goldin would not ask him too much. Instead, Goldin zeroed in on his rookie status and asked him if he thought that the mission would be safe.

Lindsey also had some stories of fun times with Dan Goldin. On multiple occasions, Goldin actually suited up and 'sat in' on shuttle launch simulations at JSC. In one instance Goldin ended up sweating profusely when the cooling system in his suit failed. Lindsey closed by noting that Goldin's tenure there had been remarkable and had included 61 Space Shuttle missions, 7 flights to Mir, and three complete expeditions aboard the ISS.

Mac Evans, the recently retired president of the Canadian Space Agency was next. He referred to Goldin as, "An inspirational leader - a true internationalist - a man of mission, passion, and conviction." He put Goldin's tenure in perspective by noting that the ISS had begun as a counter-attempt by the West to show the USSR what it could do. Over time it was Goldin, according to Evans, who turned the tables and brought the now former Soviet Union into the ISS program as full partner. Evans closed by saying that Goldin had, "Changed what humans do in space - and how they do it - forever."

Courtney Stadd, looking like he had just jumped off of an airplane (which indeed he had, having flown up to Washington after the STS-108 launch was scrubbed earlier that evening) spoke in glowing terms of his time with Goldin. He said that when he had shown up at NASA the previous December, little did he know that his tuition was about to be paid in the 'Dan Goldin School of Leadership'. He distinguished this from 'management' noting that Goldin had shown him how a great leader works—often thinking outside of the box. Stadd noted that Goldin had served three presidents and that these three 'saw the greatness in the man'. He then presented Goldin with a special plaque signed by President Bush.

One more time - with feeling

Goldin then took the podium. He flashed through the various areas of NASA's expanding involvement in new areas—noting its contributions to climate change and technology. He also took some pride in having brought the new discipline of Astrobiology into existence.

Of all the events he had experienced at NASA, there was one specific event he chose to cap off the evening, and his tenure at NASA—and it had to do with astrobiology.

Goldin recalled a time in 1996 when three scientists sat in his office for several hours. He was grilling them about discoveries regarding possible evidence of ancient life that they claimed to have made in the ALH84001 Martian meteorite. When he was done listening, Goldin was convinced that this team had enough evidence to support their claim and to go ahead with formal publication. Goldin, clearly excited, called his father.

Goldin's father was in a hospital dying of spinal cancer at the time. Goldin said that he hadn't expected him to live more than a few days.

He told his father, a biologist, of the discovery from Mars. When he was done he told his father that he'd have to keep this quiet for 8 weeks or so. His father joked with Goldin, one New Yorker to another, saying, "I am dying of cancer you moron, who am I going to tell!?"

As it happened, Goldin's father lived much longer and was wheeled in front of a television to watch the August 1996 announcement. He died shortly thereafter. Goldin, tears welling up in his eyes, said, "[That his father had] lived to see what his son had done."

And thus the evening concluded.

In December Goldin starting to settle in at the Council on Competitiveness, a frequent half-way house of sorts for senior government appointees as they re-enter the private sector. Goldin's wife Judy was often heard to say how much she hated Washington and that she and her husband would be moving back to the west coast. That did not seem to be the case.

Clearly moving into the big fish in a shrinking pond mode, Goldin once seen in public shouting into his cell phone while his secretary carried his bags. One reporter remarked that every time he called Goldin he answered the phone himself and then said, "I don't know where my secretary has gone to."

Around this time people all learned something they already knew: Dan Goldin did not use computers. He did not do email. When interviewed by the Los Angeles Times in December 2001 about his newly bought computer Goldin explained his lack of emailing as follows: "It's kind of an oxymoron. I went to NASA and went into a tech hiatus. As someone who worked for the president, I did not want to use e-mail. My conversations needed to be protected and secure in order for people to have confidence in me. I had to deal with the lives of astronauts and a variety of serious matters. I didn't want these things to be found later."

Those who worked with Goldin would say that he was concerned about many things being captured in a permanent form. He wanted nothing on the record.

As Goldin had finished packing he named Daniel Mulville to be the Acting Administrator until Sean O'Keefe had been confirmed by the Senate.

And then, after a decade at the helm of NASA, Dan Goldin was suddenly gone.

*   *   *   *   *

2

A CALL TO DUTY

The morning had begun in the West Wing in normal fashion—as if there was anything 'normal' in the post- 9-11 world at the White House. Every morning at 9:45 senior administration staffers would meet with President George W. Bush to review the latest 'domestic consequences' issues, not only homeland security matters, but other concerns as well. Just after 9:20 a.m., on Thursday 12 November 2001, Deputy OMB Director Sean O'Keefe and Deputy White House Chief of Staff Josh Bolton and others were waiting in the anteroom just off the Oval Office ("The Oval" in White House speak).

As they waited, news stations were just starting to announce that another commercial jet had crashed in New York. This time it was an American Airlines Airbus that had crashed on Long Island shortly after takeoff. It was only minutes after the crash and it was not yet clear whether this was another act of terror or an unfortunate accident. Everyone in America was still on edge at this time. Many immediately suspected it might be another terror attack. Based on these breaking events, O'Keefe and the others expected the morning's meeting with the president to be called off.

It was therefore a surprise, when at 9:40 a.m., Bush's Chief of Staff Andrew Card walked in to the anteroom and said, "The man is ready for us." As a group, they entered the Oval Office, sat down, and began briefing Bush on various issues. The meeting had just started when a phone call came for Bush from New York City Mayor Rudy Giuliani. Giuliani wanted to coordinate with the federal government agencies investigating the crash. Five minutes later, Bush was again interrupted by a call—this time from New York state Governor George Pataki, with the latest developments of how the state government was being readied in the event the plane had been downed by a terrorist's hand.

After the calls from New York had been completed, the meeting began in earnest. As had become the practice, each participant had about 15 minutes to brief Bush on current issues. Sometimes the participants had multiple items to present, other times they didn't; it depended on the day. Given the morning's latest developments, the participants hurried through the comparatively mundane issues. Bush was totally focused on the issues raised, as well as the looming prospect that terrorists had struck the American homeland once again.

When the meeting concluded the attendees filed out. The president stopped O'Keefe, saying,"Sean, I want a word with you."

Out of the blue

It had been just a few weeks earlier in late October that Vice President Dick Cheney called O'Keefe to say that the fruitless search for a new NASA Administrator was over. Sitting in his West Wing office Cheney, or the 'Veep' as staffers called him, told O'Keefe,"I've got a job for you."

There hadn't been the slightest hint that O'Keefe was going to be asked to head NASA. Nonetheless, when the President asked, O'Keefe quickly agreed to the new assignment. "If that's what he wanted me to do, then that's what I would do," he would later recall. O'Keefe wasn't just another senior administration staffer summoned back to the capital when the Republicans returned to power. He had served as President George H.W. Bush's Secretary of the Navy, tasked with cleaning up the mess resulting from the Tailhook scandal. He had been plucked out of the Senate Appropriations Committee, where he had served Alaska Senator Ted Stevens loyally. It would be a bond that O'Keefe and Stevens would maintain through all of the political jobs the new administrator would face in the years ahead. But it wasn't the OMB appointment that O'Keefe had returned to Washington to serve. He had left his academic post at Syracuse University for a possible top job at the Pentagon.

The story of who was originally proposed for what job in the new Bush Presidency is shrouded in political mythology shaped in large measure by those who ultimately won their appointments. But the cast of players at both the State Department and the building across the Potomac, that is jokingly referred to as 'having five sides and many more faces', shifted during the winter of 2001 between the end of the Florida recount and the inauguration.

Sean O'Keefe's selection to head NASA would be an echo of how Cheney, who had led Bush's search for a running mate eventually ended up with himself as the choice. O'Keefe, who had been brought in to help bring order and improved fiscal management of various federal agencies was now being tapped to head one of the most fiscally mismanaged agencies: NASA.

That bright morning, Bush himself wanted to talk with O'Keefe about NASA, in particular, about the space program. Andy Card closed the door to the Oval Office and Bush turned to O'Keefe.

"About this NASA job," Bush said, "Here's what I want you to do." Top on the president's list was stopping the hemorrhage of red ink gushing out from the International Space Station. Just days after Bush took office, Dan Goldin had dropped a bombshell on Bush's desk. A four billion dollar cost overrun had been racked up on the International Space Station's budget, and it was still growing. Goldin claimed he didn't know how it had happened, or what could be done about it. It just appeared, seemingly out of nowhere.

Now Bush gave O'Keefe his marching orders. "This place is messed up," Bush said. "I want you to straighten it out." he added. O'Keefe was told to come back and tell the president what was needed to review and to fix NASA. As O'Keefe got up to leave Bush opened the door and repeated, "Go over there and straighten this out."

While Bush had already made the decision to move O'Keefe to NASA, O'Keefe's name had previously floated about the Washington rumor circuit as being in line for another job. In Al Kamen's 7 September 2001 column in the Washington Post, Sean O'Keefe was one of two people whose names were being circulated as a possible replacement for Secretary of Defense Donald Rumsfeld, whom the media were suggesting was not up to the task of running DoD. Rumsfeld was, like the rest of the country, only days away from a sudden shock that would change everything and everyone to a wartime footing.

Curiously, O'Keefe's name would emerge again in 2004, in the very same column, this time in a much larger list of speculative replacements for Rumsfeld in the wake of the Iraq prisoner abuse disclosures. O'Keefe's name would also be tossed around in the media in coming months as a possible choice as head of the Department of Homeland Security.

One last OMB task to handle

Before the presidential encounter, just after getting the initial charge from Cheney, O'Keefe had one last gauntlet to run before word of his selection would be made known.

The occasion was testifying before the full House Science Committee on 7 November 2001. The hearing was called to examine recent decisions made by NASA in response to the Young Committee's report on the International Space Station (ISS) which had been issued just a few days earlier. After years of cost overruns Goldin had convinced Congress that all was under control. A whopping $4 billion cost overrun appeared out of nowhere and had plummeted the already beleaguered agency's reputation to an unprecedented low point. Today all the bad news and hard medicine would come home to roost.

Among the more unusual items mentioned were letters from the European Space Agency (ESA) to the House Science Committee Chair and to the U.S. State Department "expressing concern over the ISS Task Force Report" and a demarche titled 'Diplomatic Note from the Canadian Government to the U.S. State Department regarding the Space Station IMCE Report'. The cost overruns had led Young's committee to make some recommendations that struck many as being concerned primarily with U.S. issues—and that the concerns of the other nations in the program were secondary.

O'Keefe was there to present the Administration's view on the report and its implementation. The only other witness was IMCE Task force Chair Tom Young. No one from NASA appeared that day.

Even as O'Keefe sat down in room 2318 of the Rayburn House Office Building, word of Cheney's request to O'Keefe had already started to dribble out. While O'Keefe knew that he was going to be nominated, no one on the committee knew. Nor, O'Keefe thought, did anyone else.

When approached during a break early in the hearing, O'Keefe was asked to confirm word from a reliable source that he had indeed been asked to be NASA Administrator.

O'Keefe said that he could not confirm that officially but suggested that word leaking out of the impending announcement too soon might complicate things. Nonetheless within 24 hours news that O'Keefe was under consideration appeared online and speculation inside the beltway began. By now many names had made the rounds and most people treated this as just another rumor.

He was clearly focused on NASA's future as he drafted his prepared statement. Due to an impending vote, O'Keefe was urged to make his summary presentation short and sweet. However, in his prepared remarks O'Keefe said, "A most important next step - one on which the success of all these reforms hinges - is to provide new leadership for NASA and its Human Space Flight activities. NASA has been well-served by Dan Goldin. New leadership is now necessary to continue moving the ball down the field with the goal line in sight. The Administration recognizes the importance of getting the right leaders in place as soon as possible, and I am personally engaged in making sure that this happens."

There were several, somewhat pointed exchanges between O'Keefe and Rep. Dave Weldon (R-FL), an ardent supporter of NASA. Weldon had been pursuing a line of questioning wherein he described NASA as a 'whipping boy' of OMB aand he was having problems accepting some of the numbers O'Keefe had been presenting.

Weldon said, "I think it was a little deceptive the way that graph was shown. And the reason I am particularly bothered by this is, you know, you are here for the Administration and the Administration claims to be a big supporter of manned-space flight. But in the budget you all submitted, you have a two-percent increase from NASA, an agency that has been flat for 10 years, and you have got a 13.5-percent increase for NIH, an agency that has seen its funding almost double, and you are slated to increase it by another 14 billion over the next three years, up to over $30 billion. To me, it really looks like the Administration is not a supporter of the manned space flight program. I mean, when you just look at the numbers."

O'Keefe considered his reply carefully, knowing that he was soon to be named NASA administrator - but still in his OMB position - and certainly aware that he would need Weldon's help. Deflecting Weldon's question politely O'Keefe said, "I am sorry you have reached that conclusion, Congressman, but I think you have made your point very well, and I understand that, and I appreciate your comments. Thank you."

In a second round of questioning later in the hearing, Weldon asked, "Mr. O'Keefe, I have a question for you. As I understand it in previous Administrations, ever since NASA was created, the president or, in more recent Administrations, the vice president, has dictated NASA policy. The impression I get is that NASA policy is essentially being driven by OMB now in this Administration. Is that correct?"

O'Keefe replied, "No, sir. I wouldn't concur on that. I think the Vice President's interest is very strong. He has expressed himself internally within the Administration on this point."

Weldon answered back, "Can you give me a little more information on that, because I sent [the vice president] a letter. I got no response. Have you talked with him personally?" O'Keefe replied that he had and that they had discussed NASA.

Weldon had touched on a topic which many members of Congress had sensed, but few (with the exception of Rep. Nick Lampson D-TX) were willing to voice in any great detail at this point i.e. that the agency was floundering in the backwater of presidential attention - and not just the previous administration but with this one as well. The past year had been one where only bad news came out of NASA—usually relating to bad management and or cost overruns. Little good news emerged. In addition, the White House had been keeping Goldin at arm's length for nearly a year and Goldin had been doing his best invisibility act.

To top it off, the White House never mentioned space—in any context.

Few presidents ever did. As such, it was becoming increasingly difficult to remain a supporter of NASA in Congress—a stance that when taken often had a lone-wolf flavor to it.

Rumors become fact

By 13 November 2001 O'Keefe's selection was an open secret. The next day, the White House announced the impending nomination, and on 15 November President Bush formally submitted his nomination of Sean O'Keefe as the new Administrator of NASA.

Reaction to O'Keefe's nomination was optimistic but reserved. Comments from House Science Committee Chair Sherwood Boehlert (R-NY) probably best summarized the general response by Congress: "Under Dan Goldin's outstanding leadership, NASA met its technical objectives. Now, the agency needs to focus its attention more on meeting its management goals. I know Mr. O'Keefe will bring a realistic, results-oriented approach to addressing the management problems that have bedeviled NASA."

Others, scrambling to make sense of this decision, stumbled across a historical analogy: the legacy of former NASA Administrator James Webb, also a budget specialist, who was brought to NASA and ended up leading it into the Apollo age.

On 20 December 2001, Sean O'Keefe appeared at confirmation hearings before the Senate Commerce, Science and Transportation Committee. Seated to O'Keefe's left was Boehlert.

Boehlert summarized O'Keefe's capabilities and the task that lay ahead of him as he introduced O'Keefe. "Sean is a budgeteer, not a rocketeer; but he knows enough about rockets to know that they burn cash just as assuredly as they burn fuel, and that both propellants are finite. It won't hurt to have someone who can husband the agency's resources, for this is an agency that has lost its way..."

Members of the Senate committee—from both parties—voiced their support for a swift confirmation for O'Keefe such that he could be on the job at NASA as soon as possible. Both Sen. Kay Bailey Hutchinson (R-TX) and Ron Weyden (D-OR) wanted to be certain that O'Keefe would report for work right after confirmation. O'Keefe replied that he wanted to take a Christmas break with his family but, barring no problems with his confirmation, wanted to be on the job at NASA "as close to New Year's day as is possible." The only hold-up could come from a recent threat to put a 'hold' on all judicial nominations by Sen. John Kerry (D-MA) so as to force action on a small business bill he had introduced.

The senators were quick to ask O'Keefe why he thought he was qualified for the job—each asking the same question in their own way. O'Keefe was quick to note that he was not, in many people's eyes, the ideal choice. He cited his father, whom he described as a renaissance man and although retired is "in college studying astrophysics and the German language" with experience as one of the men who shaped Hyman Rickover's nuclear navy, as being a more appropriate choice. However, it was O'Keefe and not his father who was slated to head NASA. To the task at hand, O'Keefe said, "I am a public servant that has served in a range of positions. These have given me a working understanding of the management issues. My qualifications are that of a public administrator—I don't try to be chief engineer."

The hearing was not entirely a love fest. For the most part, several Senators made it very clear to O'Keefe who (they thought) was in charge. Sen. Hutchinson pushed O'Keefe to support a 7 person ISS crew-complete with a Crew Return Vehicle. In so doing she made it clear to O'Keefe that, 'You'll be my hero if you do this. But I will be all over you if you don't."

Trent Lott (R-MS) warned O'Keefe that any sort of BRAC (base closure) activity with regard to NASA's field centers would, "Meet a lot of resistance from us in the room." Sen. Bill Nelson (D-FL) made half a dozen pointed reminders that implementation of the Young Committee's recommendation to reduce the Shuttle flight rate would have a serious impact on NASA safety and the skill mix of its workforce. Nelson repeated Sen. Hutchinson's caveat, "I'll be over you," if O'Keefe were to pursue these options.

O'Keefe took these veiled threats with a composure that he would often be calling upon in the months ahead. Nodding, smiling , saying thank you politely. He wasn't in the hot seat yet—but he certainly could smell something burning rather close by.

In picking O'Keefe, the Bush Administration had taken one of their core team players and sent him on a special mission to fix NASA; to 'get back to basics' as O'Keefe repeatedly described his task. While Dan Goldin obviously enjoyed the support of three Administrations, he was never seen as being anything close to an 'insider' or a core team member. Indeed many in these three Administrations found interactions with Goldin to be difficult and sometimes uncomfortable.

O'Keefe, on the other hand, was a consummate insider. Indeed, to extend the analogy, O'Keefe had helped write the very playbook NASA would now be expected to adhere to—with O'Keefe as head coach.

When coming into a chaotic situation where a controversial leader is being replaced, there is a temptation to blame the other guy for things. Sen. Nelson would say, "When you put [things] in a context that NASA has been a 'bad boy' that is when I get concerned. No less a space giant, than Chris Kraft, has written a letter about the Young Committee Report. He found it difficult to understand that such a group would present such a narrow view. You would think that members of a task force would try to understand ISS history - that today's financial status was preordained." Nelson said, "I don't want to just focus on what is wrong—but to fix it. But I don't want to use this to punish NASA." The Senator's observations carried with them much credibility.

In what would become standard practice for O'Keefe in dealing with problems that plagued NASA, O'Keefe replied, "Since January 20th the president has directed us to view all matters as looking forward—not the past. There is no intent to punish. My intent is not to try to unearth what led to this circumstance—it is WHERE WE ARE. We need to get back to basics and define requirements. I want to build on what is there and utilize existing capability that is there." In the coming months this would be refined into what O'Keefe often derided as 'archaeology" - looking back and dwelling on who did what - and when. Rather he dipped into the past only to get data he needed to move forward.

Late night arm twisting

Early in the morning of 21 December 2001 the Senate confirmed O'Keefe's appointment. It almost did not happen. This was the last night of the legislative session and the Senate was about to adjourn for the holidays. Sen. Kerry had placed a hold on all nominations for about 10 days . The net result was that O'Keefe and other nominees (and senior military personnel awaiting promotions) were more or less held in limbo until the Administration agreed to let Kerry's bill move ahead— from Kerry's perspective, that is. Senators John McCain (R-AZ) and Ted Stevens (R-AK) worked on Kerry, but Kerry wanted an assurance from the White House before letting O'Keefe's confirmation (or any others) through. Eventually, White House Chief of Staff Andy Card had to intercede and passed on an assurance to Kerry that the Administration would give the bill a fair look. McCain tracked Kerry down on an airplane and got him to release his hold on the pending nominations. As such, the path for the Senate to vote on the nominations was now clear.

Just after midnight the Senate voted to confirm O'Keefe, the new Director of BLM, several U.S. Attorneys, and a swarm of army generals (many of whose promotions had been held up for some time) —and did so about 30 minutes before the Senate adjourned. Had this last minute intervention with Kerry by McCain, Stevens, and Card not been orchestrated, O'Keefe (and the others) would have had to wait a month for the approval of their nominations until the Senate reconvened.

With legislative approval now under his belt, all of O'Keefe's official paperwork was prepared and sent to the White House on 23 December 2001. The president drafted a short note accepting O'Keefe's resignation as Deputy OMB Director and then signed the order appointing O'Keefe as NASA Administrator. That step completed, O'Kccfe was officially sworn in by David Addington, the Vice President's General Counsel (and a Hill pal of O'Keefe's) at a small gathering in the Old Executive Office Building.

Showing up for work

Sean O'Keefe, NASA's newly minted Administrator, fresh from an actual week's vacation, showed up for work bright and early on 2 January 2002. At this point he still preferred to make the drive-in under his own power from his home in Suburban Virginia. Wasting no time, he issued a personal note to all NASA employees. In so doing he sought to begin his tenure looking forward—not backward:

"As I move into this new role as NASA Administrator, we face a substantial 'to do list'. It's going to require a lot of hard work and some difficult decisions. But with you, I know we will reinvigorate the agency's mission of discovery and conquer new challenges.

"NASA leads a unique expedition that is vital to the future security and vitality of our nation and humanity.

"As we celebrate this season of renewal, let us resolve to face the problems, step up to the challenges, exploit the opportunities, and continue to pioneer the frontiers of air, space and knowledge. In 1899, Theodore Roosevelt said, "Far better it is to dare mighty things, to win glorious triumphs, even though checkered by failure, than to take rank with those poor spirits who neither enjoy much nor suffer much, because they live in the gray twilight that knows not victory nor defeat."

Facing the enemy

Barely on the job for 4 complete days O'Keefe did something his predecessor would have avoided at all costs: he held the first of two morning breakfasts with 20 or so reporters. Held in the large meeting room on the 9th floor of NASA Headquarters, the first event was held on 8 January 2002. The following day a second group of reporters was invited in for breakfast and some quality time with O'Keefe.

O'Keefe ran the sessions himself in a relaxed, cordial manner—one that bespeaks someone who is comfortable with himself—and with the job. In those rare events when Dan Goldin was publicly available to the press his press secretary Glenn Mahone would manage the event.

O'Keefe did not need a handler and picked reporters on his own. Unlike Goldin whose mannerisms in such a venue would be stiff and dripping with ego and suspicion, O'Keefe was remarkably at ease and cordial.

He didn't make much news at these events. Much of what he said was recycled from his confirmation hearing. But that was not the point. Being accessible was. All comments were on the record, indeed, O'Keefe's policy with regard to things he says and writes is that everything is always on the record. This way, he reasoned, "You never have to stop and wonder whether you actually said something." The logic being if you are always thinking that everything you say is going to be under scrutiny you tend to put more thought into what you say.

Some concern had already been expressed in the media that Sean O'Keefe was not the ideal choice for NASA Administrator by virtue of not having an overtly space-oriented background. The general impression of all of his predecessors was that they were all affirmed space aficionados.

When asked if he had a 'burning passion' to be NASA Administrator, O'Keefe said that he had a passion that, "Whatever you do - you do it right." He went on to say that, "There are few Americans who do not have an appreciation for the enormous capability that this organization has. It is very easy to get excited about this." He was then asked if this was a 'dream job' for him. He shot back instantly, "You bet. I am just astonished that the President had the confidence in me."

The remainder of the questions focused about the topic du jour in those days: cost problems on the ISS program and how the agency was going to dig itself out from under this looming crisis. O'Keefe, clearly still learning the ropes - and the jargon - tried to steer clear of overt pessimism - or optimism - saying instead that the problems would need to be examined closely and worked through. His newness to the job and the topic showed - and he did not shy away from the obvious.

However, just as O'Keefe was putting his toes in the water for the first time with the NASA press corps, indications of a disconnect vis à vis space policy emerged across town.

Singing from different song sheets

A few hours after O'Keefe met with reporters, presidential Science Advisor John Marburger addressed attendees at the Annual meeting of the American Astronomical Society. A few minutes later he held a press conference for reporters. In marked contrast to O'Keefe's cautious course regarding space station problems, Marburger presented a far more gloomy potential future for the ISS.

When asked to comment on the Bush administration's views on robotic versus human exploration of space, Marburger said, "We need to review what we get out of investment we get in space-based research. There is something to be gained from having people on the spot to do things.

"The Space Station is up and running and has people on it and represents a great investment by the American people. It would be a great scandal if it were not exploited."

When asked if he felt that the U.S was risking violation of a treaty level commitment with regard to issues relating to the international partners, Marburger said, "I don't think it will come to that. We value the partnership and the science. We are in a very tense situation trying to get control. The incoming management [at NASA] is competent and good. The Young report has a lot of good things in it."

Marburger then said, "There may be a greater chance of abandoning" the ISS. And then he quickly said that, "This is not the place (in front of several dozen reporters) to say this. Don't quote me on that." Moments later Marburger said, "If we cannot get our arms around the management problems in the Space Station much greater dangers could lie ahead."

This disconnect would continue for another year as O'Keefe struggled to drag NASA back to respectability while facing the reality that his agency - and its charter -were nowhere near the top of the administration's priorities.

Let's get it right this time

While he had taken the formal oath, in private, weeks before, O'Keefe was publicly sworn in as NASA Administrator by Vice President Cheney on 15 January. The ceremony was held in the Langley Theater at the National Air and Space Museum in Washington. 450 people were in attendance. A reception in the museum's exhibit area beneath a swarm of rockets followed. Both events were by invitation only and involved heavy security procedures; standard fare in post 9-11 Washington. Among the more lively attendees was O'Keefe's father who made a point of talking up his son's qualifications, and his pride in them, to anyone who would listen.

The attendees were a who's who of government and industry. Among those acknowledged from the podium as being present were Deputy Secretary of Defense Paul Wolfowitz and Deputy Secretary of State Richard Armitage.

Vice President Cheney hailed Dan Goldin for his work at NASA, but was unabashed in his enthusiasm for long time friend O'Keefe, saying, "The President and I believe

Sean O'Keefe is the best man for NASA at this time. The President and I believe in NASA and stand behind this agency ... we are confident in its future and its ability to do great things for our country."

Cheney then said, "This is the fourth time I've sworn in Sean O'Keefe... we are determined that we will continue to do so until he gets it right - and we think that this might well be that time!"

O'Keefe thanked Cheney and said that he would not 'let you and the president down,' and that he saw NASA as "an important part of the national security community ... I will give increased attention to that aspect of the agency's role in the federal government." He added that he was "astonished that I have been given this dream assignment" and that his son Jonathan chided him, saying, "That's great, Dad, but I thought you had to be really smart to get that job. You know, you're no rocket scientist."

O'Keefe added that he was "astonished that I have been given this dream assignment" and that his son Jonathan chided him, saying, "That's great, Dad, but I thought you had to be really smart to get that job. You know, you're no rocket scientist."

Road trip

O'Keefe wasted little time in meeting his workforce. He started by visiting Langley Research Center on 10 January 2002. He would think back later on the visit as amazing, for what it told him about the state of the agency he had inherited. Very specific instructions had been given by Courtney Stadd to the attendees.

The Enterprise Associate Administrators were to make a package of presentations lasting three or four hours. Soon after O'Keefe and his party arrived, Sen. John Warner (R-VA) showed up, to see if O'Keefe planned to close down the center. O'Keefe assured Warner that he wasn't planning any such move. But as soon as the presentations started, it quickly became clear none of the presenters were going to follow Stadd's script. In fact, the Associate Administrators talked all right about themselves, not about the agency. It turned out to be a poor exchange of information. O'Keefe listened dispassionately and only made a few remarks.

Afterwards the exchange fueled O'Keefe's desire for a housecleaning. And a housecleaning it would take. By the time of the latest retreat in the spring of 2004, 85% of the people who had been in his senior leadership team in early 2002 would be gone. What was left was a drastically new team.

O'Keefe was on the road again for Johnson Space Center on 17 January 2002. While at JSC he was introduced by House Whip Rep. Tom DeLay (R-TX) who had urged O'Keefe to visit the Houston center. DeLay (who had avoided visiting JSC for years) introduced O'Keefe and laid it on thick: "Here at the Johnson Space Center, we're closest to the passion that captured America's imagination. I'm speaking of the talent and tenacity that sent men to the moon and brought them home safely." In so doing DeLay touched on an aspect of NASA culture that would dog O'Keefe every day: the fiefdoms carved out over the decades by each center—each one asserting that they were the true heart for one aspect or another of NASA's various programs when, in reality, such passion was to be found all over the agency.

The road show then reached Marshall Space Flight Center on 22 January, Glenn Research Center on 25 January, Dryden Flight Research Center and Ames Research center on 30 January. The remaining centers were visited within the first weeks of February.

O'Keefe's style emerges

Shortly after this trip, O'Keefe began to make his first significant impressions on the way the agency would operate. The occasion was the new membership of the NASA Advisory Council (NAC). What caught many by surprise was how the new administrator would reach utterly outside the standard pool of nominees. Moreover, the ties to the Navy (O'Keefe was a Navy brat and former Navy Secretary) and to O'Keefe's academic haunt, Syracuse University, would soon become familiar aspects of how he sought out advice.

Among the new names was a very familiar one: former senator and astronaut John Glenn. Given his long career in the senate and two flights in space, no one would think to question his qualifications to serve on the NAC.

The remainder of O'Keefe's nominees, while certainly prominent in their own right, were not at all familiar to the NASA family. Indeed, the ties that O'Keefe had to them, and the prevailing naval theme that ran through their backgrounds, left many scratching their heads. Just what was this guy up to?

The new members included Richard Danzig, Secretary of the Navy from 1998 to 2001 and Undersecretary of the Navy from 1993 to 1997; Roger E. Tetrault, retired vice-chairman and chief executive officer of McDermott International (O'Keefe had served on the board of McDermott). Tetrault was also a graduate of the U.S. Naval Academy and once served as corporate vice-president and president of General Dynamics' Electric Boat Division, manufacturer of the bulk of America's submarine fleet; Dr. Donald C. Fraser who served the George H. Bush administration beginning in October 1991 as the Principal Deputy Undersecretary of Defense for Acquisition (while O'Keefe was still serving as DoD Comptroller and Chief Financial Officer); and David J. Berteau, Director of National Security Studies (O'Keefe's previous position) at Syracuse University's Maxwell School of Citizenship and Public Affairs, the U.S. Department of Defense Executive Management Development and Training Program, in New York.

Over time many more people from the Navy, Syracuse University, and from O'Keefe's previous jobs would appear on NASA's doorstep as advisors and employees. While a Navy bias would often roll off of some people's lips, no one ever went so far as to suggest anything unprofessional or unethical was behind this. They would, however, grumble every time 'another navy guy' was hired on.

Rather, as many would soon learn, Sean O'Keefe had a career wherein he managed to develop a long list of 'friends' who, by nature of how O'Keefe came to know them, also had expertise that was relevant to what O'Keefe was trying to do at NASA. This marked one of the starkest contrasts to Dan Goldin who arrived in Washington as an outsider, and, despite his decade at NASA's helm, remained an outsider, never developing many true friends. Although people would often say they 'respected' Goldin, this was simply an acknowledgement that Goldin managed to get his way.

There was one appointment in particular that O'Keefe made that was overtly personal in its genesis. One of the most influential appointments O'Keefe made was that of long time college pal Paul Pastorek as NASA's General Counsel. Less than a month after O'Keefe arrived at NASA word surfaced that Pastorek, a partner in the law firm Adams and Reese LLP was under consideration for the NASA General Counsel's position. Pastorek is an alumnus of Loyola University and got his B.A. in 1976 and his J.D. in 1979. Sean O'Keefe got his B.A. from Loyola in 1977.

Pastorek and O'Keefe were long time friends. The two were roommates in college and the two earned extra money together doing sheetrock installation and other construction tasks. While Pastorek would indeed perform the role of being NASA's lawyer, his day to day role was much more pervasive. Pastorek, a frequent after hours companion of O'Keefe's, would exert an influence in all aspects of policy. Indeed, the role Pastorek played at Sean O'Keefe's NASA was not at all unlike the role Robert Kennedy played when his brother was president. This relationship would deepen in the months ahead, and would form the basis of O'Keefe's most crucial assistant in the dark days for NASA that lay ahead.

O'Keefe saw the selection of his inner circle as his most important task. While not formally defined by role or responsibility, who was in this inner circle would soon emerge as O'Keefe got to know the people around him. An early meeting of the minds - and admission to this inner circle - happened between O'Keefe and another tall Irish-American, Bill Readdy. Very close in age, and outlook this relationship would be fortified by adversity barely a year later.

An example of how O'Keefe dealt with people was the way that Bill Readdy found out formally that he was going to be the Associate Administrator for the Office of Spaceflight. O'Keefe, Readdy and several others had adjourned to a favorite bar of O'Keefe's in New Orleans. At one point O'Keefe said to Readdy, "I feel awful about this. There is some paper work I need you to read over and sign-off on." As Readdy went through the small pile of paper he came to a press release requiring his approval. The press release announced Readdy's appointment as Associate Administrator for the Office of Space Flight.

Everyone had a good laugh.

The first setback

O'Keefe was not without his setbacks. Barely five days after the Senate voted to confirm him as Administrator, O'Keefe's choice for Deputy Administrator was already circulating: former astronaut and Marine Major General Charles ("Charlie") Bolden. Universally admired and genuinely liked across the agency, Bolden's consideration added instant credibility to O'Keefe. NASA workers were heard to say, "If this was the caliber of people he'd have at his right side, he must know what he is doing."

On 31 January 2002 the White House formally announced Bolden's nomination. Less than two weeks later, Jefferson Davis Howell Jr. was named by O'Keefe to replace Roy Estes as Johnson Space Center Director. At the time sources close to Howell, a retired U.S. Marine Corps Lieutenant General, said that he had accepted the position knowing that his friend Bolden would be at NASA headquarters.

But Bolden's nomination would never get to the Senate.

In early March 2002, Defense Secretary Rumsfeld was grappling with an agency that was now on a war footing in Afghanistan. Short of personnel in many areas he sought to reverse the flow of military personnel to billets outside the agency. In a 6 March 2002 press conference Rumsfeld said, "As your question suggests it creates an enormous incentive for the defense establishment to stop using the men and women in the armed forces for things other than military tasks. We have them spread all over town as detailees, in one place and another."

Rumsfeld and Sean O'Keefe, while avowing mutual respect for one another, did not get on-at all. Indeed some of OMB's decisions on Rumsfeld's initiatives had made things worse. With no love lost between them, Rumsfeld was not about to have someone at NASA overrule him. In addition to Rumsfeld's stance, Bolden ran the risk of falling out of the promotion queue if he took off for NASA. While many sources suggest that Bolden would have taken the job at NASA—career implications and all—O'Keefe said to people in private, "I could not ask him to do such a thing." O'Keefe was truly impressed with Bolden and really regretted that he would not be able to call on him to help.

As such, on 13 March 2002, just hours before he was to go up to Capitol Hill for his confirmation hearing, NASA announced that Bolden's name had been withdrawn by the White House.

O'Keefe sought to put a positive public spin on this by saying, "Senior military leaders of General Bolden's caliber are a rare and precious resource. The Marines are very fortunate to be able to keep him among their ranks. Given the ongoing war on terrorism and the imperative expressed by the Secretary of Defense that all uniformed military personnel serve to advance the president's objectives to win the war, we fully support the president's decision." Behind the scenes O'Keefe was furious, stunned, and deeply disappointed by being outmaneuvered by Rumsfeld.

In May the name of Frederick Gregory was announced as O'Keefe's new choice to be his deputy. Gregory had been moved from being Associate Administrator of NASA's Safety office to being the Deputy Associate Administrator for the Office of Space Flight and then, with the retirement of Joe Rothenberg, he was named as the Associate Administrator. Gregory was widely known, well-liked, and was confirmed promptly and without any fanfare.

Dueling fiefdoms

One of the things that caught O'Keefe's attention as he entered the agency was the extreme amount of fractionation that existed in the agency's organization—due mostly to the ten field centers it had spread across the country. Each one often acted as if the others did not exist. Indeed, a visit to their websites would leave a visitor hard pressed to find mention, much less links to other field centers—or to NASA headquarters itself.

O'Keefe was prompting a notion he called 'OneNASA' which sought to pull down the barriers between centers as well as curb the duplication of effort (often called 'stovepipes') wherein multiple centers would try and keep a foothold in a variety of areas. A cynical workforce, having been pounded by a decade of Dan Goldin's management fads, was openly skeptical.

Few NASA employees would, however, disagree that a cohesive, unified and reinvigorated NASA would be a wonderful thing to have. After a decade of a Machiavelli at NASA's helm, having a Don Quixote might be an effective change—even if he didn't know how a windmill works.

By now, after decades of existence, no shortage of advisory reports, exploration scenarios and the like were imprinted on people's minds. Now that the leadership vacuum had been filled, everyone now seemed to want to know 'where' NASA was going to go next i.e. what celestial destination.

To many, O'Keefe, an avowed space novice, was a blank slate everyone now sought to sway. Still up to his knees in budget issues and holding to every option, O'Keefe did not want to go down the path of citing specific destinations—especially given that the one orbiting overhead was so troubled.

The traditional approach in formulating a space policy was to pick a destination or endpoint and then let that designation pull the technology along with it. That is how Apollo was done. O'Keefe reversed the polarity on this old argument and said that it would be premature to pick a new destination until such obstacles as travel time were reduced. Then, and only then, O'Keefe felt, would NASA be in a position to make a choice of destinations.

O'Keefe would stick to this mantra rigidly for the next year. Only when Columbia was lost and a pervasive re-examination of America's space policy began would he relent and see the value of picking a destination and then retooling his agency to meet the challenge.

But that was still a year away—on the other side of a national catharsis. At this point, O'Keefe was still feeling his way down a path blazed by what he often referred to as 'this storied agency'. Clearly he'd have to lean on others to set the agency back on a path; but he would soon surprise many as to how fast he learned the ropes.

Education and visions

Something that would catch the eyes of some close observers of O'Keefe was his interest in education. With two children in junior high school and one in high school, he was clearly a parent who saw much of the world through his children's eyes. This would quickly manifest itself in his charge to the agency in the Spring of 2001 and his first major announcement.

O'Keefe returned home to Syracuse, to the Maxwell School of Citizenship and Public Affairs at Syracuse University to announce that Barbara Morgan, astronaut and former Idaho schoolteacher who served as Christa McAuliffe's back up, would fly in space on a space shuttle mission.

This announcement ended a 15 year limbo for Morgan. The announcement was couched in a larger context: O'Keefe would announce that a new crop of 'Educator Astronauts' would be selected. However, unlike the last round in the 1980's these selectees would be trained as full fledged astronauts, as Morgan had now become.

This program was then set against an even larger backdrop: a change to the agency itself. A month or so after O'Keefe arrived at NASA internal discussions began as to how to frame a new vision for the agency—one that would fit on a single piece of paper and be easy enough for people to recall. Previous NASA 'vision' documents tended to be wordy and jargon laden and somewhat confusing to outside readers.

In a speech titled 'Pioneering the Future', O'Keefe announced:

The new NASA vision for the future is:

> *To improve life here*
> *To extend life to there*
> *To find life beyond*

> *The NASA mission is:*

> *To understand and protect our home planet*
> *To explore the Universe and search for life*
> *To inspire the next generation of explorers . . . as only NASA can*

Of course, something this simple, while having the benefit of getting a simple message to a wide audience left many within the agency confused as to if it related to them or whether this referred to another part of the agency.

It would also leave Congress a bit baffled as well.

*   *   *   *   *

3

FIELDS OF DEBRIS

"Exactly what do you want me to do?"

It was a crisp fall day, at least as far as what passed for fall in New Orleans. Paul Pastorek was working, as usual, at his desk at Adams and Reese, a large law firm downtown. His specialty was mergers and acquisitions, which was exactly what was occupying his energies and attention when the call came.

It was his old friend Sean O'Keefe, calling from Washington. Pastorek had celebrated, along with O'Keefe, when George W. Bush had won the presidency in the December 2000 Supreme Court decision on the Florida recount. He had celebrated even more when O'Keefe told him Vice President Cheney, another in O'Keefe's wide circle of friends and mentors, had summoned him back to federal service in the new, second Bush administration. It had looked, for a time, like O'Keefe might be headed across the river to the Pentagon for a prestigious post.

But instead, O'Keefe had arrived at the Office of Management and Budget as its deputy director, a position that Pastorek thought his bean-counting pal would love. So it was something of a surprise when O'Keefe came on the line and told Pastorek he wanted to see him the next time he headed north.

Nothing urgent, O'Keefe had said.

"About what ?" Pastorek asked.

"About something," was all O'Keefe would say on the telephone. His friend suggested that sooner would be better than later, but whenever. Pastorek was then managing the Washington office of his firm, so a visit to the capital wasn't out of the ordinary.

It was in November 2001 that Pastorek would fly up and meet with O'Keefe at one of his favorite watering holes, the Dubliner, across from Washington's Union Station. O'Keefe had frequented the Irish establishment since his college days, and had come to know personally the owner and his family. They settled in and began a round of icy beers.

O'Keefe said that President Bush had summoned him to the Oval Office without warning, and had asked him to serve as NASA Administrator. "I had to make a decision on the spot," O'Keefe told him. He had accepted, and then went home to tell his wife that night he would be moving from OMB to the space agency. It would be announced to the public in a few days. O'Keefe then asked his friend, "I'd like you to join me at NASA as General Counsel."

Pastorek, who had never so much as given space flight a second thought, was stunned. "I asked him 'what does that job pay?'" he recalled later, and didn't much care for the salary O'Keefe reported. He joked that much of his 'former comfortable' lifestyle would be crushed in the process. It didn't appear that he could afford to accept the job, but was intrigued in any event. "Exactly what do you want me to do?" he asked.

O'Keefe said there were two major issues that loomed ahead as he prepared to take over control of NASA from Dan Goldin. One was the space agency's contract with the Jet Propulsion Laboratory (JPL). The other was the Space Flight Operations Contract, or SFOC as it was called inside the space program. United Space Alliance was a corporate firm formed by Boeing and Lockheed Martin to operate the fleet of NASA's space shuttles, under the conditions of the SFOC agreement. He wanted him to take those contracts apart and in the deconstruction understand them fully.

"I didn't know what those (acronyms) meant at the time," Pastorek recalled. But he would be forced to learn - and fast.

And there would be other issues to address as well, mainly for Pastorek to serve as O'Keefe's main enforcer, change agent, and closest associate in a sprawling bureaucracy where he knew virtually no one. O'Keefe also wanted him to revamp the space agency's education programs. Pastorek had worked on the Louisiana State Board of Education; O'Keefe knew his friend would bring that background as well to NASA. But would he take the job? "I said I would think about it, but it seemed to be a real non-starter on the financial side."

"It would require a big change in my life," Pastorek recalled. "I really wasn't interested in moving to DC." To add to the consternation his daughter was still in high school. But he thought about it some more, talked it over with his law firm, and by December it appeared that he would be headed to NASA, first as a consultant 'while they ran the traps', and then as a presidential appointee. In early January, 2002 the White House, having been satisfied that Pastorek was sufficiently Republican, signed off on his assignment.

None too impressed

His first exposure to the NASA 'culture' was the first leadership retreat ordered by O'Keefe, the journey to Langley Research Center in Virginia. It was there that both O'Keefe and Pastorek came to agree that a new management team was sorely needed, "The need for financial rigor, and management rigor in that agency," he recalled. "My impression was that Sean was none too impressed."

Pastorek managed to avoid talking to most people, and mainly listened. He was shocked by what he saw and heard from the space agency's 'best and brightest'. "It didn't look like a very homogeneous organization that had a clear vision about where it was headed," Pastorek remembered. "It looked like there was a lot of work to be done." While the lack of clarity in the presentations surprised him, the extent of NASA's ills didn't. He had been briefed in advance by O'Keefe as to what to expect. After all, the place had a $4 billion cost overrun on the space station project, and that might be just the tip of the iceberg.

Change, he thought. Change was badly needed in the place.

And, as they settled in, that's precisely what O'Keefe began to do.

Loss of comm

Those events and actions would be but a prelude to what was to come. And when the day came, no one in the top leadership of NASA saw it coming.

It was a 'bluebird' day, as pretty a morning as one could ever hope to see on the east coast of Florida.

Paul Pastorek looked out onto the sunlit runway and saw nothing. And heard nothing. He, like everyone else gathered at Kennedy Space Center the morning of the first of February 2003 was waiting for the arrival of Space Shuttle Columbia. But he wasn't alarmed by not hearing or seeing anything. In fact, he wasn't even supposed to have been there at all. Pastorek had flown east to Kennedy Space Center from New Orleans. He arrived late, and didn't see any NASA people until the next morning.

NASA's leaders, including Sean O'Keefe, Kennedy Space Center director Roy Bridges, and Associate Administrator for Space Flight Bill Readdy clambered aboard a bus for the ride to the runway. Pastorek had his rental car from the night before, and he drove behind the bus as they all headed for the shuttle landing strip. His plan was to leave the car at the strip and return with O'Keefe aboard the NASA jet back to Washington. The following Monday the Fiscal Year 2004 budget submission was to be released, and Pastorek wanted to go through the briefing one more time with O'Keefe as they flew home.

That was his plan, as he milled around the site where VIP bleachers for the families and for other space dignitaries had been erected. Pastorek had bought a new digital camera and was trying it out as they awaited Columbia's arrival. "I took about three pictures; Readdy and Roy talking, the families in the background," he remembered. As he played with his new toy, he remembered hearing in the background concern for clouds. There were clouds headed towards the landing site, and since the shuttle was a glider in its final phase aloft, could not detour away from any deteriorating weather conditions.

Then Pastorek heard that communications with Columbia had been lost. He didn't see that as a big deal. "I'm not knowledgeable enough at that point to realize that after TDRSS you don't lose comm," he observed. But he looked over at Readdy and Roy Bridges, who were quietly talking amongst themselves. He looked over at O'Keefe, who was holding forth in a lively conversation with someone that he had once worked with. O'Keefe was oblivious to any concern—since no one had yet informed him of what was starting to unfold. "He didn't seem anxious."

As O'Keefe broke off the chat and strolled over to Pastorek, Readdy came up to them, his face white. "Something is wrong, terribly wrong," he told them.

Pastorek was startled. He and O'Keefe looked at each other in disbelief. Pastorek turned to Readdy and suggested maybe there was an explanation for the communications interruption. "Maybe they landed somewhere else," he said. "Maybe they landed in the water?" "I mean", Pastorek suggested, "doesn't this happen all the time?" Readdy said no. "No, Paul, we should have heard the sonic booms by now." Well then, Pastorek suggested, "couldn't it just be late?"

And Readdy, his face drained of blood, told him no. "It always happens precisely at a specific moment." Now O'Keefe began to show his concern. Pastorek again suggested that maybe Columbia had landed off course. And he, and O'Keefe, were chilled by Readdy's reply.

"We just don't know where it is."

O'Keefe asked Readdy, "Well, what do we do now?"

"Sean," Readdy said, "you're going to have to declare a contingency." He pulled a binder from his ever-present briefcase, and opened its pages. "These are the things that are going to have to happen," he told O'Keefe, starting with moving the astronaut's families out. "At that point, Roy was starting to give the instructions on moving the families out."

A sudden shock hit O'Keefe. One of the first things O'Keefe did when he came to NASA was to order that an overhaul and review be made of contingency scenarios such as this one. He had never imagined that he'd have to actually live through one.

Now functioning in automatic mode, O'Keefe, Readdy, Pastorek and Roy Bridges all piled into Bridges' car and headed off to the nearby Launch Control Center (LCC). As they drove off, Pastorek could see the astronaut's families being hustled into a van. When they arrived at the LCC, they were quickly taken to a suite of rooms upstairs, on the opposite side of the building from the control operation.

In one of the empty offices, the senior management of the civil space program waited. O'Keefe had called the White House during the five-minute drive from the landing strip. People were now on their cell phones, or on office phones, attempting to understand what was happening in the confusion. O'Keefe was trying to get hold of Andy Card, the president's Chief of Staff, and reached him. Card said that Bush was at Camp David, and he would notify him of what was happening.

"We had a foam strike."

"I was calling legal people," Pastorek said. Others were calling the Federal Emergency Management Agency (FEMA). A high-level phone conversation with FEMA officials was arranged for later in the morning. Ron Dittemore, shuttle program manager, came on the phone for O'Keefe. O'Keefe asked Dittemore what was going on, what had happened to Columbia? The CNN broadcasts that they were seeing on a TV in the office indicated that something had happened, but it wasn't yet clear what it was.

Dittemore's answer shocked O'Keefe. "It looked like a problem with the left wing," Dittemore told O'Keefe. He then said that it looked like, "We had a debris strike on the left wing during the launch," and that it appeared that it had happened in the same place where they had registered off-scale readings from instruments embedded in the wing just before contact with Columbia had been lost. Dittemore said that he didn't know if that was the cause or not, but, "We did have a strike on the wing."

Pastorek looked over at O'Keefe, who had no idea what Dittemore was talking about- a debris strike on Columbia's wing? As the conversation with Dittemore was taking place, someone walked into the room and handed Pastorek a color photo, the kind printed out from a computer. It showed a debris strike hitting Columbia seconds after its lift-off on 16 January 2003. No sooner had Pastorek and O'Keefe been handed the photo, they noticed under the CNN broadcast a crawl that said, "A debris strike had occurred on the left wing during the launch."

Pastorek was stunned that the media knew as much as they did. "Ron, how the hell do these people know that?" Dittemore told him that nobody at NASA was telling them that, or anything. "This was the first hint that everything that happened at NASA was going to be on TV before we knew about it, including the administrator." It was the first time O'Keefe or Pastorek had ever heard about debris hitting Columbia during launch.

Fields of Debris

Still reeling from what he had heard, Pastorek looked over to the TV tuned to CNN. Somebody had taped videos of what appeared to be flaming debris, "About five or six red balls of fire coming down." He motioned to O'Keefe to look at the television images. As it became obvious that what they were seeing was the destruction of Columbia, the room fell silent. Pastorek grabbed the phone with Dittemore still on the line. "Where the hell is that coming from?" he demanded to know. "Is that our imagery? Whose imagery is that?" Dittemore answered by telling him that wasn't NASA's imagery of Columbia falling from the skies. "We don't have any imagery," Dittemore told him.

"We were flabbergasted," Pastorek recalled as they saw the balls of fire streaming across the TV screen. "Then the horror really sunk in. And we didn't know how or why." By then, O'Keefe had spoken with President Bush several times. The president's biggest concern was the families. Bush asked what NASA was doing to take care of the families. How are they doing, he wanted to know. "It would be a theme, a clarion call that resounded over and over again. We are going to do everything in our power for the families," Pastorek recalled.

O'Keefe then brought his senior leadership into an hour-long meeting. Pastorek remembers that there were about 150 people in attendance. The subject was how the space agency was going to deal with the crisis, mainly from an engineering perspective. "How are we going to investigate? Who would do the clean-up?"

While O'Keefe listened to the options his team were developing, a call came in from FEMA. Pastorek grabbed him and took him to the phone. They talked, Pastorek recounted later, about who was going to be in charge of the clean-up, and it was agreed that FEMA would take the lead role, and NASA would be its partner.

O'Keefe and Pastorek returned to the meeting. As they took their seats, Pastorek looked over to the TV set. It was now showing a crawl that said, "A press conference at NASA coming up." O'Keefe, who was sitting at the opposite end of the room, saw it too and motioned for Pastorek to come over. Pointing to the screen, he asked Pastorek, "I'm just looking at the screen here and it says we're going to have a press conference in 30 minutes. Do you know anything about this?" he asked. "I said I have no idea." Roy Bridges was sitting next to O'Keefe, and said nothing. The administrator wasn't happy. "Go find out who called this press conference," O'Keefe demanded.

Mass Confusion

A call was placed to Associate Administrator for Public Affairs Glenn Mahone, who was organizing the NASA media response from the agency's E Street headquarters. He didn't know anything about a press conference. Pastorek then asked that he call the public affairs team at Johnson Space Center, to see if they had called it. While JSC was being polled, technicians were attempting to set up a secure video link with Bush at Camp David. He wanted to speak directly to the Columbia families. But a secure line couldn't be established, so Bush wound up speaking on a telephone link-up.

O'Keefe turned to his staff. "We're not having any press conference until after the President speaks," he told them. The president spoke gently with them for about a half hour, saying that he and the First Lady were praying for them and their lost loved ones. And Bush said something else. That they would get to the bottom of what happened that morning.

As the President came off the line, the press conference issue kept coming up again. Mahone had called back from Washington to say Johnson Space Center didn't plan anything either-and that they wanted to know what they were supposed to tell the press in any event. Reporters were clamoring for information, and there was precious little to release.

Finally, Pastorek had enough of the time wasted on the whole matter. "We're not having any press conference," he told O'Keefe. They had nothing to report, and no one knew who had ordered it. "I turned to Roy (Bridges) and said, 'Roy, you don't know anything about any press conference, do you?'." Bridges spoke up. Yes, he said he had called the press conference. Pastorek exploded. "Why?" Bridges said that according to the contingency procedures plan, it was at this point following an accident that the NASA Administrator was supposed to face the press. "We don't want to make the same mistake we did in Challenger," Bridges told O'Keefe.

Pastorek then closed the issue. "We're not holding any damn press conference until we know what we are going to say—and who's going to say it."

The families

O'Keefe, Pastorek, and Bill Readdy then went in to see the families. On the way to the holding room, O'Keefe asked Pastorek about what he should be saying to the media and the public. "Go draft me something," he asked his friend, and Pastorek peeled off and went into another room to start writing.

In the meantime confusion still reigned. Glenn Mahone was also writing remarks for O'Keefe to give the country. But they had been lost somewhere in the NASA Kennedy Space Center bureaucracy. Mahone had faxed them, but nobody seemed to know where the fax machine he sent it to was located. After about 45 minutes, O'Keefe emerged from the meeting with the families. It had been, he would recall much later, among the most difficult meetings he had ever conducted as NASA Administrator or in any other part of his professional life. He took the hand-written speech Pastorek had written, and started working on it, making slight changes here and there.

O'Keefe, Pastorek, and Readdy jumped back into Bridges' car and headed to the press site to face the nation. Pastorek called Mahone and instructed him to tell the press O'Keefe was on his way. As they headed to the press center, another car pulled up alongside and signaled frantically for them to stop. Pastorek wondered if this was more bad news. Since they had first seen the TV pictures of the flaming debris falling from the sky, they had steeled themselves for the news of injuries to people on the ground. But none had come. Now someone was flagging them down.

Bridges pulled the car off the road and Pastorek rolled the window down. A KSC person handed over Mahone's missing speech, which had been found after all. But O'Keefe was more comfortable with Pastorek's text, which he had already altered to fit his tone. He passed the Mahone speech to Readdy. "Here Bill," O'Keefe said, "why don't you make use of this."

A word to the world

The group entered the press center. After a brief delay to organize themselves, O'Keefe headed to the door that led to the auditorium. But Pastorek noticed there was a commotion. People and TV camera crews were blocking the entrance. There in the midst of the crowd was Florida Senator Bill Nelson. Pastorek saw Nelson say something to O'Keefe, which caused him to respond. Then, loudly, Nelson said, "Mr. Administrator, you're going to earn your spurs on this one!" Nelson then turned and smiled at the camera.

Pastorek was fuming. They had been through the worst disaster to strike the agency in the history of NASA. Seven people were dead, a multi-billion dollar spaceship, the first of its kind, had been torn apart descending from orbit, all on live television. Nearly an hour long meeting with the families left everyone drained and in tears.

Out of the doorway the national and international media were waiting to hear O'Keefe speak for the first time since Columbia fell. And now his boss had to be confronted by this spectacle. "It was unsettling to say the least," he recalled. "I thought that it was flat-out inappropriate," Pastorek said later.

It was about 4:00 p.m. when O'Keefe finally mounted the stage. Many U.S. networks, and all of the nation's 24/7 cable TV channels now switched live. "This is indeed a tragic day for the NASA family, for the families of the astronauts who flew on STS-107, and likewise is tragic for the nation," he said. O'Keefe explained that immediately after the shuttle had failed to appear at its landing time of 9:16 a.m., he had activated Bill Readdy's contingency plan.

O'Keefe had contacted President Bush and Homeland Security Secretary Tom Ridge, briefly advising them that they had lost contact with Columbia, and with its crew. Bush offered O'Keefe his full and immediate support to determine that appropriate steps be taken. O'Keefe told the press he had spent the next 90 minutes working through the details and the information that had been received, addressing operational and technical issues. He said Readdy would give those details after he completed his briefing.

"We have to act responsibly"

O'Keefe said that he had met with the Columbia family members and that President Bush had spoken with them as well. Bush, he said, had expressed his regrets, and had pledged to them that the process would begin immediately to recover their loved ones and understand the cause of the tragedy. At this point, O'Keefe said that there had been no indications that anything on the ground had contributed to the accident. Citing the contingency plan, he said that a mishap investigation team was formed at 9:30 a.m. that morning, and that it was coordinating its work with a rapid response team from Kennedy Space Center and gathering together the data from the lost spacecraft from Johnson Space Center. The other centers involved in the shuttle program were also contributing to the effort.

These activities, internal to his agency, were about to be supplemented. An accident investigation board with an external group of people would be formed, and it would be independent of NASA. Moreover, it would be charged with the responsibility to look at all of the information which had been immediately locked down right after communications from Columbia had stopped. O'Keefe described the board's members as safety and mission assurance people from other federal agencies, and from the Air Force and Navy. He did not say who the members would be, and who would lead their investigation. He also described the effort to band together other federal departments and local and state agencies. NASA's role to them, he said, was to clearly define Columbia's reentry path from space, across west Texas towards Florida. He was concerned that fragments and wreckage from the shuttle be secured on the ground where they had landed.

O'Keefe had a special concern about people coming into contact with the materials. "We would urge people who believe they have found any material to stay away from it, and please contact local officials," he warned. He announced that FEMA was coordinating the recovery task, along with local first responders in the area where Columbia's remains had rained down. For the most part, it appeared that the majority of the terrain where the shuttle parts had impacted was rural countryside. There had been no reports of property damage yet, or injury to people.

And then he closed with a message about accountability. "We diligently dedicate ourselves every single day to assuring these things don't occur," he said. "And when they do we have to act responsibly, accountably, and that is exactly what we will do. The loss of this valued crew is something that we will never be able to get over." He hailed the astronauts as people that had dedicated their lives to pushing scientific challenges—and he asked that the news media respect the families' privacy, and to "please understand the tragedy that they are going through at this time."

When O'Keefe and Pastorek returned to Roy Bridges' offices on the grounds of Kennedy Space Center, they were met by a group of congressmen and senators, including Bill Nelson. O'Keefe moved off into another room to talk with them. For the most part, they all offered their help and support for him, and for NASA, during the investigation. In the meantime, Pastorek was in telephone contact with Deputy Administrator Fred Gregory and Glenn Mahone in Washington. The recovery of the debris had started, so organizing that didn't seem to be a problem for NASA. FEMA seemed to be well up to the task. "Johnson Space Center was running with the ball on that," he said. One by one, technical or political issues arose and they were decided, on the spot.

But one recovery decision loomed ahead. And that was what would NASA tell the media?

"The truth doesn't get better with age."

There were mixed feelings about how to handle that issue. Pastorek went through a 'devil's advocate' process. "I like to consider every option," he explained. "Stupid or not." So now he addressed the issue. What would they tell the news media? And how would they tell it? What was the consequence of each approach? Or of doing something else? Anticipating what's going to happen "in three hours, five hours, as opposed to handling the problem that's on the table at the moment."

"We have options," he explained to the group. NASA could take the path of full openness. Or it could take a legalistic approach, and basically say 'no comment' every time. Or it could take an approach that lay between those two extremes. They argued back, and forth, several times. Several participants changed their views or took the position of others.

Finally, when there seemed to be no consensus, Bill Readdy spoke up again, "I think we should take the approach that my father taught me," he said. "And that's to tell the truth, tell it all, and tell it now."

O'Keefe listened to Readdy, as he had listened to the others. This decision would shape NASA's response to the Columbia accident for the rest of the investigation. It would establish an image that not just the country, but the Congress would interpret as symbolic of how NASA saw itself, and its ultimate responsibility for what had just happened. They did remember the days following the Challenger's loss in 1986. Then NASA did everything it could to avoid telling the truth until the last minute. The result, many believed, tarnished the space agency's image, and clouded a full appreciation of the shuttle program, for years thereafter.

O'Keefe spoke. "The truth doesn't get any better with age." NASA would tell it all, as it was learned, as it was uncovered. Good or bad. And they would have to live with the consequences.

Pastorek then warned that there would be consequences for that openness. "There is going to be some document somewhere," he predicted. "I have no idea what they are." But his legal experience told him that there would be documents that they would discover that would be embarrassing, to those in that room and to the space agency in general. "So I flat out asked them - is that the kind of stuff you'd want to disclose?" O'Keefe's answer was 'yes'. "We want to disclose sooner rather than later," O'Keefe said. They all went through a litany of different types of scenarios, and how to handle the media or the political fallout in each case. And there was also another consensus.

The making of the CAIB

"There was agreement that the one thing we would not do publicly was discuss what might have been the cause of the accident. We're not going to get involved in that."

O'Keefe's reasoning was obvious. They had spent much time during the day's discussions on the formation of what would be called the Columbia Accident Investigation Board (CAIB). And the discovery of what caused Columbia's destruction would be their job, not NASA's.

O'Keefe had several ideas for the types of people who should be on that panel. He was suggesting positions, not names as yet. He had some names for potential chairmen of the panel. He picked up the phone and called Fred Gregory. "Fred, here's some people I want you to call concerning this accident board," he told Gregory. He also suggested that Fred consider some names as well. It turned out that Gregory had been thinking of names since ten that morning. The name of retired Admiral Hal Gehman was mentioned by someone. Gehman had been in charge of the investigation into the attack on the U.S.S. Cole. He asked that several of the people call O'Keefe or Gregory back. One of them was Gehman.

No speculation

O'Keefe began to think about the investigation, and NASA's role in it. "If we sit here and speculate about what did or did not happen, we're not going to be pleased," O'Keefe said. He reasoned that if NASA concluded it was caused by some thing or another, his agency would be perceived as having a vested interest in the outcome of the investigation. "Whether we do or not."

After all, if there was this independent board making the assessment of the cause, O'Keefe said let them make their assessment. He called Ron Dittemore back, and warned him that he could say anything he wanted—except speculation on potential causes of the disaster. "Do not speculate on the cause of the accident," he was told. You can talk about what happened, you can talk about how foam struck it, and what NASA did or didn't do in response to it. "Good, bad or indifferent," Pastorek recalls Dittemore being told. "Talk about all that stuff. Just don't say 'I think this caused it', or 'I think this didn't cause it.'" That, Dittemore was told, isn't your province.

Briefings

That evening, O'Keefe, Pastorek, and Readdy boarded the NASA jet for the trip back to Washington. It was a somber flight, as the craft took off, shortly around 9:00 p.m., and sped north. At one point during the flight home, O'Keefe got up out of his seat and came over to Pastorek and spoke in a quiet voice. "The world as we know it has changed," he said. "If we do not do this right, we could lose human spaceflight, for good." They were pondering the implications of that as they landed at Reagan National Airport at 11:00 p.m..

Such was the pervasive opinion at NASA: lose another shuttle after Challenger and the game would be over. O'Keefe would find himself battling this decade-old unofficial NASA prediction of what might happen if another Shuttle and its crew were to be lost. He was not so sure that he totally agreed with this dire prediction. Indeed, there might be a path out of this darkness.

Among their early tasks as they all returned to NASA Headquarters was the briefing schedule for the accident, before the investigation board came together. Readdy and Pastorek, with input from Glenn Mahone and others, decided on a plan. Each day, twin briefings would be conducted. One at NASA Headquarters and the other at Johnson. It would be a detailed airing of the progress in gathering the data from Columbia's last seconds and the recovery and search for wreckage.

A morning session would start the day off in Washington, followed by Dittemore briefing the press in the afternoon. The plan was to do these briefings from NASA, beginning that Monday February 3rd all the way until Friday, when the board would take that function over.

The plan worked well, until Thursday's session at the Johnson Space Center. During what would have been Dittemore's next-to-last press session, he mentioned, almost in passing, that the debris strikes might have caused the accident. Pastorek was watching the broadcast on NASA Select, the space agency's own TV channel. He grabbed the phone and called O'Keefe, whose office was down the hall on the opposite end of the 9th floor from Pastorek. "Did you hear what I just heard," he asked O'Keefe. "That wasn't in the script, was it ?" Now the very thing that they had wanted to avoid had been said. In Dittemore's defense, all he really said was, "You are going to be getting something that shows you're going to see a strike on the left wing, and there may be a correlation or maybe not." But now a foam strike as a leading accident cause scenario was out there.

A minimal mass

O'Keefe then proceeded to misspeak himself, this time before Congress. He had received two pieces of information from NASA's Michael Greenfield, which indicated that a foam impact on the orbiter's wing would do little damage. The early analysis was a rough calculation about the speed the foam object would be moving, at the point it struck Columbia's wing. O'Keefe was told it was just a few hundred miles an hour. And that's what he said at the Congressional hearing. He also said, based on an early analysis, that the magnitude of the foam's mass would be minimal. And he said that in public, too.

"I cringed when he made that comment," Pastorek recalls, but it turned out that both statistics were way off from the reality of what Columbia experienced during launch. It took fluid dynamics to reveal the actual truth of what had happened that morning. And what had happened was fatal, although the actual mass and impact speeds would take months to be fully understood. And even then, the foam as the cause of the accident appeared to be unlikely.

"We were all speculating internally as to what it might be," Pastorek said, recalling that period in the early months following the accident and as the CAIB got underway with its investigation. No one theory seemed adequate, until the foam issue became better understood. NASA's engineers were working alongside the CAIB team. Hal Gehman had been selected as the board's head after all, and he moved quickly to gather the data radioed down from Columbia, as well as look at the debris that started to pile up from Texas and Louisiana.

"No, thank you"

Attending the memorials and funerals for the astronauts was emotionally hard to absorb. "Those were the most gut-wrenching things that I've ever been to," Pastorek remembered. "I had never cried, and Sean had never cried so much as a consequence at some event." But both men, and Readdy as well, also made trips to Texas to review the debris recovery. There were teams of many people from many different agencies, and walks of life. He recalled native Americans out on the searches, scouring the countryside for traces of Columbia.

At night, there would be camps of maybe 500 or 600 people living in tents. And he recalled there was one big tent under which everyone ate dinner at the end of the day's searches. There was a makeshift shower. At one location there was a building nearby that served as an auditorium. There, O'Keefe would give a presentation on the status of the recovery and the investigation, firefighters and others would explain what they were doing. Pastorek remembers that O'Keefe always took an astronaut or two with him. And at the end of his presentation on NASA's activities, he had the astronauts sit down and sign autographs for the folks there. Some nights it took hours for the astronauts to sign all of the photos and things that people handed them.

And then something interesting happened, a shift, he thought.

On one trip, as they were waiting for the astronauts to complete their autographs, someone came up to O'Keefe with a shuttle picture. O'Keefe was asked for his autograph. "Next thing you know, someone else came up, and another, and another." One night, when O'Keefe was besieged with autograph requests, it took almost three hours for him to sign everything. O'Keefe was amazed. But these were people that were proud to be a part of the shuttle program and the space program. It was real to them. And they thought they were helping to be part of the effort to return America to flight. To one man, O'Keefe said 'thank-you'. But the man replied 'no, thank-you for letting us be a part of this.'

Helping the families

The other issue was working with the families as the recovery wore on. O'Keefe would meet with the family members once a week over the first six months following the accident. Sometimes he flew to Texas to see them. He had asked them to allow Pastorek to attend these meetings, saying, "He's a lawyer but also my friend. Someone to bounce ideas off of." But the Columbia families would not agree. They told O'Keefe to keep Pastorek in Washington.

Sometimes, NASA dropped the ball, in regard to some issues with the families. Sometimes issues weren't implemented quickly enough. Sometimes NASA rotated the astronaut assigned to them out after he was assigned a flight. This occurred with astronaut Andy Thomas, who had developed a strong relationship with the Columbia families. When he was rotated out due to a mission assignment, several objected. They thought Thomas had been removed because he did something wrong, or because NASA was mad at him. As it happens, the agency had neglected to warn the families and their friends that Thomas was bound for a shuttle mission assignment, when the flights resumed and he needed to get on with his training. O'Keefe had to go and personally explain it, apologize for the oversight. "What we have tried to do, within the law is address their concerns," Pastorek said. Any cash benefit or award that could be given on behalf of the lost crew to the families was devised.

And generally, as the year wore on, the relationship, unlike that of the Challenger families with NASA, remained strong.

And also unlike the Challenger accident, there were no lawsuits.

Shattered image

As the CAIB investigation wore on, NASA's reputation slowly deteriorated. As Pastorek had predicted on the day of the accident, documents, in the form of emails and studies done by Boeing and USA, had suggested the foam strike might well be more threatening than NASA's safety managers thought. Worse yet, requests to image Columbia in orbit by national security satellites were denied repeatedly. Such pictures might have revealed the unsuspected damage on the spacecraft's left wing, caused, as the board would soon conclude, by the very foam that was the immediate cause that people thought of in the minutes following the disaster.

Equally astonishing, the lessons of the 1986 Challenger accident had been forgotten by the NASA that Sean O'Keefe inherited. The sobering realization would hit O'Keefe and his team again and again that Spring, almost like a physical body blow.

From the memorials and funerals, to the debris collection, Congressional testimony and the drum-beat of bad news from the CAIB investigations, O'Keefe would be hammered relentlessly. As NASA's head, that was his job, of course. But this wasn't the role he had expected to play when President Bush summoned him to lead the space program. This wasn't about exploring the solar system, or building space stations.

This was about death. Needless, avoidable, gut wrenching death.

If those in his charge had only listened to the spacecraft, they would have understood its warnings. Columbia, like Challenger's o-rings before it, had been giving clues to its foam-strike problems almost since the first launch in 1981. But NASA's safety system wasn't listening.

The whole disaster, as O'Keefe was learning, could have been avoided.

Coping with the implications

Long after the cause was understood, after his agency had returned to something akin to normal days the focus became, not Columbia, but getting her sister shuttle Discovery ready and capable of returning the shuttle fleet to service. But the loss of STS-107 continued to haunt Sean O'Keefe and his band of reformers. It would follow them like an unseen shadow. "We were responsible," O'Keefe would think. "We killed those people." And sometimes, not often but sometimes, the energy of his determination that NASA never diminish the loss of that shuttle and crew came to the surface with emotion.

One such occasion came in the spring of 2004, long after the accident and the recovery had faded from the news. O'Keefe came before the American Astronomical Society, intending to talk about the space agency's future, advanced telescopes and science missions. But he inevitably had to speak, again, of his rationale for canceling the fifth space shuttle servicing mission that had been planned for the Hubble Space Telescope.

He went through the litany of the path his reasoning took as he arrived at the decision. And then he joked about the press attacks his decision had triggered. "Journalists have written stories about other journalists and the empowered opinionated who are offering this view, and describe the criticism as 'withering'," he said. He called it pretty much standard fare for commentary, "On just about everything around Washington these days," he joked again.

And then his face froze and he turned somber and serious. "Instead, let me offer my view of 'withering'," he said. Withering, he coldly explained, "Is the feeling you get when you are standing at a runway with the dawning realization that the shuttle everyone is waiting for isn't going to land." "Withering," he went on, "is when you have to explain to wives, husbands, parents, brothers, sisters and children that their loved ones aren't coming home alive. Withering is attending funerals, memorial services, and ceremonies over 16 months in number too many to count anymore." And then he paused, "Yet having every single one of these events feel like the weight of that responsibility will never be relieved." "Withering," he ended his retort, "is the knowledge that we contributed to the Columbia disaster because we weren't diligent."

And at the end of his speech, he departed from his text to talk, again, of the accident. "I am wearing the STS-107 pin," he said, grabbing his lapel. "I wear it each and every day; I can't imagine not wearing it. Because we must always be diligent and make certain that we never forget, that we never have this happen again." And then looking up at his audience, "As each of you leave here today, there is a STS-107 pin for everyone that wants one."

His soul was seared on that runway, and in the days, weeks and months that followed. But others, equally affected by the disaster, found new missions for the lost shuttle Columbia.

Keeper of the relics

One that did was NASA employee Scott Thurston. Each day, Thurston arrived at the Kennedy Space Center and helped prepare orbiter Atlantis for her role in the return to flight preparations. Atlantis would be kept on a stand-by basis to launch in a rescue mission if Discovery suffered the kind of damage that befell sister ship Columbia.

Thurston's career, in a sense, was bound up in the Columbia accident. He had been in charge of preparing the Columbia for its fatal, final flight. Thurston, and a NASA and United Space Alliance team had poured over the winged ship, installing payloads, and readying its systems for its flight. "We had a lot of science interfaces," he recalled many months after the accident. "It was fun, because we hadn't done that in a while." The Columbia, the oldest of NASA's flying shuttles, needed a bit more 'care and feeding' from Thurston and his team for each of its missions. Typically, he recalled, once Columbia was in orbit it flew with little problem. But its quirky personality was missed. "We don't treat the orbiters as machines, per say," he said. "We call them 'shes'."

And so it was that Scott Thurston, too, was waiting on that runway on 1 February 2003 for the spaceship that would never return. His team was to take possession of Columbia, remove its flight hardware and get it ready for its next space assignment. But instead, the next day he received orders to head to Texas. There, for the next four months, instead of preparing Columbia for space flight he would help retrieve its shattered pieces.

The biggest pieces that Thurston and his team would find were about 1 foot square sections of reinforced carbon-carbon, as well as thermal protection system tiles. There was no trouble identifying the materials as being parts of a spaceship. "We touch this hardware every day, and look at it, so it was very obvious it was orbiter hardware," he recalled. The broken parts had the telltale signature of high reentry heating effects as well, making the pieces even more obvious. Although, there was at last one piece that was identified that didn't come from the shuttle. An elderly lady arrived at Thurston's camp in Corsicana and said she had found part of the shuttle. She handed over her prize. It was a bowling ball with the name 'Columbia' painted on it.

But the whole retrieval and recovery exercise was almost like a catharsis for Thurston and the other technical reps that had flown down from Kennedy Space Center to, as he would say, "Help bring her home."

After the debris had been cataloged and used by the accident investigation board, the question remained: what to do with the wreckage? NASA managers tapped Thurston, and asked if he would study whether or not there was any scientific interest in keeping access to the Columbia wreckage open. Thurston reviewed the kind of storage space that would be needed, and what the cost might be. After all, there were 84,000 pieces that had been retrieved, about 84,000 pounds of Columbia. By weight, that amounted to about 38 percent of the 220,000 pound winged spacecraft. He studied all of the potential storage options; from entombing Columbia in the same missile silo that held Challenger's wreckage, to renting space off the grounds of KSC, to finding a place, somewhere, on the base. He figured about 10,000 square feet of air-conditioned storage space was needed.

NASA eventually released a formal Request For Information to potential researchers. Was there something that Columbia could help address? To their surprise, NASA received back about 20 serious research ideas, and 16 formal research proposals. NASA told Thurston to prepare to lay out the debris, in yet a new field. What emerged as the best location for the task turned out to be, literally, right in front of everyone's noses.

In addition, to spending part of each day getting Atlantis ready for the return to flight, Thurston sometimes detoured over to the vast Vehicle Assembly Building. The gigantic structure was built to house and assemble the huge Saturn V moon rockets, and had been retrofitted to accommodate the smaller but more complex shuttles and their tanks and boosters.

Arriving at the foot of the security gate, Thurston would flash his badge, then produce a key to a side elevator. Only one of three people with such keys, Scott would turn the key and ride to the 16th floor of the building. There, when the elevator doors opened, he would turn and unlock a door that enclosed 7,000 square foot of refrigerated space, the entire length of the 16th floor.

It turns out that when NASA built the VAB for Apollo, it had built in space to house some 3,000 people in offices. Long since abandoned as office space, the area that he would enter contained row upon row of archival boxes. Lining one wall, the boxes contained small objects carefully wrapped in bubble plastic, indexed, cataloged, and entered in a computer database. It was the wreckage of Columbia. Next to the wall of boxes was the larger items recovered from Texas and Louisiana, such as windshields, parts of the wing, reconstructed models from the accident investigators, and rocket engines and landing gears.

But it was the objects that lined the other wall that Thurston said often reduced visitors to tears. There, all along the far wall NASA and USA employees had lovingly preserved the hundreds of cards, letters and posters sent in by people from all over the world, expressing loss for the Columbia and its crew, saying prayers and wishing NASA hope and recovery from the dark disaster. The largest posters hung from the ceiling, flapping in the air-conditioned breeze each time its tranquillity was disturbed by a visitor, or a research request. In such a case, Scott would unseal the room, turn on the lights, find the desired piece of debris, and make an entry in both a visitor's log and the wreckage database.

Scott Thurston considered this to be Columbia's latest mission, and his along with it. After all, when he had started with NASA 14 years ago, his first mission was STS-32, flown by the Columbia.

That said, each time he needed to enter the room, he would do so with great care. Then, he would quietly retrace his steps, reseal the door, turn out the lights, and once again plunge Space Shuttle Columbia into a quiet and darkened field of debris.

*   *   *   *   *

4

CAIB AND ACCOUNTABILITY

"Strategery"

As the time for the release of the CAIB report drew near, the halls of NASA were filled with speculation as to what harsh judgments it would contain. "There were clear signals that were coming to us as to what might be in the report," Paul Pastorek remembered. "We scored all of that." In truth, NASA had been carefully following the hearings and since its own engineers were the ones working with the board during the tests on foam strikes and analysis of the Columbia's data transmissions and wreckage recovery, the direction the board would take was increasingly obvious. Since the senior managers had a strong flavor of what would be in the report, Sean O'Keefe ordered the development of a fundamental plan that would lay out the space agency's reply.

Already, shock waves had rolled over NASA as to what had come out during the spring and summer hearings on Capitol Hill. "There was no *there* - there in the safety program," Pastorek recalled CAIB chairman Hal Gehman testifying. "That was the worst of the worst. It just blistered a whole lot of people around here - I mean, big time blister." And they didn't like to be blistered, either. To say they were unhappy was an understatement, and some were spoiling for a fight. Confront the accusers head on. Defend the agency. To hell with Congress. NASA people in safety organizations were livid, Pastorek said.

In the aftermath of Gehman's remarks, and other elements of the investigation's conclusions that had been made public, Pastorek drew the senior leaders together under O'Keefe's direction to hammer out their formal response. Would NASA seek to agree? Or, as some were suggesting, would the agency take the gloves off, and for the first time since February 1st take its critics head on?

"We knew that they were going to make recommendations, and we knew that they were going to make findings," Pastorek said. "Findings of fact were conclusions of law—they are factual." But what would be their conclusions as to what had to change?

So a few days before the report's release, the debates continued. O'Keefe ordered Pastorek to go and write a preliminary statement, a draft of what the agency would say publicly. He did, and with 48 hours to go before the report, he presented his work before O'Keefe and the senior leadership, a regular group whose deliberations were jokingly called the 'strategery' group after the phrase coined on 'Saturday Night Live' during the 2000 presidential campaign.

His statement shocked some, for without knowing what fixes to the shuttle or to NASA that CAIB would recommend, Pastorek's statement said, "We are going to accept the findings and comply with the report." Sight unseen, before they knew what the implications of such an embrace would be.

Not surprisingly, the draft caused a 'tumult' in the leadership. Some objected, saying, "We can't accept these things without reading what they would be!" And O'Keefe and Pastorek understood those concerns, and they weren't really being unreasonable. After all, if this were a trial, how could a lawyer handling the case say he agreed with the judge's verdict if he hadn't even seen it first?

Accept and comply?

Such a reaction would be a normal response, but Pastorek argued that it wasn't a normal situation. And the debate formed two arcs of concern. One was over the phrase *'accept the findings'*. The other, the more contentious debate, addressed this idea of saying *'comply with the recommendations'*. Accepting the findings, O'Keefe was hearing, was easier to swallow than the recommendations, which could conceivably have serious consequences to the agency's future—and that of the shuttle.

The findings would likely say that there was nothing in the safety program. Some were already objecting to that prospect, as they had howled when Gehman said as much during the hearings. "Sean was having conversations with Gehman." Pastorek recalled. But there were no leaks from the board to NASA. No inside information. "We were doing what everybody else was doing," he said, which amounted to a careful reading of tea leaves. And there were many tea leaves to read. The board's public hearings, statements, and press reports were combed through carefully.

O'Keefe, Mahone, Readdy, and Pastorek had maintained a good inside agency intelligence process of analyzing the tenor of what was being said about the accident, as well as the Hill reactions. Much of that was public bluster, with an unseen effort to reach out to NASA. Despite their public utterances, many on Capitol Hill wanted to see NASA survive the disaster, and possibly made stronger. And despite the public humiliation and depth of the Columbia tragedy, with all of its potential impact to the shuttle program, the shuttle enjoyed strong, if not bipartisan support.

Pastorek had read and summarized all of the documents that had been sent to the CAIB, and read and summarized the damning emails from the NASA, Boeing, and USA workers who were concerned about the foam strike effects before the day of the disaster. It was an indictment, by fact. "Factually, you did something wrong," Pastorek said. "All of that would be in the report's findings. The recommendations were going to be 'here's what you have to do to correct it.'"

Would they kill the shuttle?

Pastorek's critics felt it was bad enough the board was going to indict NASA, and now he wanted the agency to just 'wholesale agree' to whatever it was? That was hard for a lot of people to accept.

Many were worried that Gehman would recommend that the shuttle should never fly again, that it was unsafe. Such a recommendation would be another disaster for NASA, for the space station components that were sitting in storage at the Kennedy Space Center were awaiting a launch aboard the shuttles. In addition there was unfinished hardware around the world, specifically designed for launch aboard the shuttle. Without the shuttles all this space hardware would be marooned—destined to become expensive museum pieces.

But O'Keefe's careful reading of Gehman's statements, and his conversations with him about the Columbia and about shuttle safety, hinted that no such recommendation was likely. "He (Gehman) had telegraphed openly that he wouldn't recommend that," said Pastorek.

Yet there were other concerns—that the report would so tightly bind NASA into specific actions and proscribe specific results from the actions, that it could actually create a new safety problem. "We thought, 'Be careful, Mr. Gehman, if you say you have to do this, this, this and this—and then you'll be safe.' If you do that, you'll take all of the interest in our part to be safe out of how we do things," Pastorek suggested.

Sean O'Keefe had an understanding about the form of the recommendations fairly early during the investigation. The internal debate raged in the final hours before the release. Finally, after everyone had had their 'rant', O'Keefe spoke. NASA would accept their findings, and comply with the recommendations, however hard they might be.

Dinner with Hal's team

The CAIB's final report was therefore neither a complete surprise nor a warm embrace for the agency. In O'Keefe's view it was, as it had been all along, important to embrace it—whatever the costs—and make its negative assessments the basis to move on to shuttle safety reforms. But defining the pathway to get to those reforms, and to both absorb the CAIB recommendations as well as head off any more political problems it might create, would prove much harder than he had expected.

On the evening of the day that the report was released, O'Keefe had a two-hour session reviewing its findings with nearly all of the CAIB members present. A few were missing, Roger Tetrault, Douglas Osheroff and Sally Ride had prior commitments and couldn't be present. There was no snub, O'Keefe felt, just scheduling issues.

After the meeting, Pastorek escorted the group to the Columbia Café on the 9th floor of NASA Headquarters for a quick dinner. He would recall much later the atmosphere of the dinner as being cordial with no finger pointing or 'incidents'. As the dinner drew to a close, it was agreed by Hal Gehman and O'Keefe that it would be fruitful if some of the CAIB members would occasionally return to meet with senior NASA leaders to chew over the report in greater detail. After all, O'Keefe said, it would take time to digest the full series of its recommendations; some of which hadn't yet sunk in.

O'Keefe's leadership team would need to read the report over, think about it, and discuss it. Inevitably, questions would arise. Gehman agreed, but also told O'Keefe that the final report document was the 'definitive word' from the board. There would be no further embellishments needed nor should any be expected. Any follow-on sessions that he would agree to have with NASA would only serve as a way for NASA to obtain explanations on what the CAIB had found and recommended and to get advice on how the recommendations could be implemented. Gehman and O'Keefe agreed to those ground rules, and the dinner concluded.

Policing the board

Several weeks later, such a meeting for clarification of the report's observations and findings between O'Keefe and some of the CAIB panel members took place. After it ended, to continue the dialog and to also continue to foster cordiality, O'Keefe invited several board members to continue to chat over dinner. This time dinner was a more formal event in his own private conference room adjacent to his office at NASA headquarters. Hal Gehman came again, as did CAIB members John Logsdon and Steven Wallace. As they sat down, they were all joined by NASA General Counsel and O'Keefe confidant Paul Pastorek.

Logsdon decided to deliver a lecture—its subject 'culture'—and the Navy/Marine Corps model as one that NASA should emulate. O'Keefe, being a former Navy

Secretary, knew something about this, and started to pepper the professor with questions, probing to see what he really knew about the subject and where he had learned it.

Slowly, it dawned on O'Keefe that Logsdon had no real depth or knowledge about that which he was speaking about, and in fact, he was 'blissfully ignorant'. Listening to his answers, O'Keefe thought, "He doesn't know what the hell he's talking about." Logsdon portrayed the Navy submarine and naval reactor community as indicative of the broader Navy/Marine Corps culture.

While O'Keefe thought that there was much to admire in the way that part of the sea service did its business, to suggest that its practices were widely adopted throughout the services didn't reflect the reality he knew. "This is just pure fantasy," O'Keefe thought as he listened to the academic. O'Keefe turned to Gehman, who had 35 years of experience in the navy, and later, as joint warfare commander. But instead of injecting his own views or experience, Gehman let Logsdon's 'errant description' stand. "Hal Knows better," O'Keefe thought, "Logsdon's excuse is that he doesn't." As the discussion wore on, O'Keefe's temper gradually rose. He responded to some of the professor's comments, each time looking over to Gehman for support. But to no avail. For whatever reason, O'Keefe found himself becoming enraged.

Rather than cause a scene or confront the old teacher, he excused himself and walked the short distance to his office. There he sat trying to cool off when Pastorek walked in. "What's wrong, Sean?" he asked. Whereupon O'Keefe unloaded on his friend, complaining about Gehman's silence and reluctance to confront Logsdon on even a simple matter as the culture discussion. After O'Keefe had finished spouting off, Pastorek gathered him up and the two rejoined the remains of the dinner.

When it was over—'an unceremonious end'—O'Keefe went back to his office to cool off. He turned the incident over in his mind. It was Gehman that he was really angry at, not Logsdon. It gradually dawned on O'Keefe that he couldn't count on Gehman to police his own board members, now in the glare of publicity after the report's release. Indeed, they could say anything they liked about NASA, no matter how far off the mark it might be, and Gehman wasn't going to correct them.

O'Keefe had watched Gehman's face that night. He was clearly tired, in O'Keefe's phrase, of 'herding cats', and had worked in the final weeks during which the report was written to keep them in line. Now that it was over, Gehman wasn't about to try and rein them in again. As a result, O'Keefe thought, some members like Osheroff were constantly feeding the press statements—statements that the member neglected to warn O'Keefe were coming.

The result was that NASA continued to get pounded in the press, like a million cuts of a knife. Some of these folks enjoy reading their names in the paper, he thought, and Gehman has tolerated all of it. Now, it became clear to O'Keefe that he would have to deal with each CAIB member on an individual basis. He thought back to the night of the report's release, and their pledge that the report itself would serve as the last word on the subject. "I guess," he thought, that, "they hadn't heard the last word after all."

A full embrace?

One evening after the report's release, Sean O'Keefe, Bill Readdy and Paul Pastorek sat around the administrator's office and talked about the report. Pastorek said NASA now needed to go one step further. It needed to say 'we embrace this report'. Readdy wouldn't hear of it. "We've already said we will comply, I don't see why that is necessary," Readdy said. O'Keefe and Readdy went round and round over that issue. There was a standoff, and Pastorek said later, "Well, I got nowhere with that." But the next day, O'Keefe had a press event. And used, for the first time in public, the phrase, "We embrace this report."

Pastorek didn't raise the issue again. And when Readdy spoke to the press the next time, he, too said, "We embrace this report."

Slowly, as the weeks passed, the phrase cropped up in more and more NASA public utterances. We accept the findings, we will comply with the recommendations, we embrace this report. "For the first 30 days after (the CAIB report release) it was a continuous refrain," Pastorek said. "We had to make sure we come clean about our mistakes, and that our people were prepared for the hearings." For to say that NASA 'embraced' the CAIB report, suggested that there had been "a pretty thoughtful understanding of what you did wrong."

But Pastorek was also mindful that you couldn't walk around all the time and admit that you make mistakes. But when you make them, especially on the scale of the Columbia accident, you have to admit it, as a first step in the process of correcting them, Pastorek said.

That long internal struggle, and public response, ended a phase of the Columbia accident that had opened on February first. Gehman's report had said that the shuttle wasn't unsafe, but needed a lot of change, both physically and culturally, to take to space once again. Now the hard work would start of fixing the shuttles and make them safe again.

While NASA was feeling the sting from the CAIB's report, and the consequences that

Pastorek had warned would inevitably come from full openness, O'Keefe's predecessor once again was in the news.

Please don't come to work, Dan

Just as Dan Goldin seemed to have faded off to a quiet post-NASA career, he managed to make news again. In so doing he did it in a way that confirmed just about everything everyone who had every worked for him had ever believed. It soon became clear as people read news reports from Boston that not a thing had changed about the way Goldin affected those with whom he interacted.

After leaving NASA, one of Dan Goldin's business affiliations was with Cassidy & Associates, a lobbying and consulting firm located in Washington DC. Jeff Lawrence, who served as Associate Administrator for Legislative Affairs under Goldin, was a Senior Vice President at this firm for a while. Among his many accomplishments, the firm's founder, Gerald Cassidy, was elected to be a member of the Board of Trustees of Boston University.

In July 2003 Boston newspapers began to publish stories about Dan Goldin being considered for the job as president of Boston University. Within a month, after nonstop negotiations with the University, Goldin had been offered and had accepted the job. Most folks involved had apparently not known much about Goldin and had been impressed with his intelligence, tenacity and the career he allowed them to see. But as his interactions with them grew, more of his character began to show.

As the initial dance was occurring, the CAIB issued its report. While it did not lavish a lot of attention on Goldin, it did paint the actions undertaken within the agency he was responsible for running for a decade as being seriously flawed. Among the flaws found were cuts in safety budgets.

Again, the CAIB did not vilify Goldin, but it did manage to encapsulate his attitude by noting a comment he had made in 1994. "When I ask for the budget to be cut, I'm told it's going to impact safety on the space shuttle. I think that's a bunch of crap." Such comments made the rounds in the Boston papers—and throughout NASA.

Within a few weeks the initial enthusiasm for Goldin began to sour. By October the Boston University Board of Trustees was ready to vote and reconsider the offer it had already made to Goldin. It would seem that many had re-evaluated Goldin's 'temperament' (a word often used) to be president of Boston University.

As for Goldin's way of doing things, little had changed. The Boston Globe reported in late October 2003, "For example, Goldin told some trustees that he had asked a psychiatrist to analyze [Chancellor John] Silber, and the psychiatrist concluded that Silber was a 'paranoid megalomaniac', according to two sources close to trustees."

Feisty and combative as ever, The New York Times reported, "[Goldin] said that he would be ready if the trustees wanted a fight." "Please know that I am advised by my personal attorney and by distinguished Boston counsel I have recently hired as to my rights," he wrote. "I take those rights very seriously and will protect them vigorously." He also reminded the trustees that he had a 'binding contract'.

Things only got worse and soon came to a head. According to the Boston Globe, "One day before he was to take office, former NASA chief Daniel S. Goldin backed out as the next president of Boston University after clashing with long-time president and chancellor John Silber."

Goldin had reached an agreement with the University wherein he received a substantial financial settlement. The Globe characterized the settlement like this, "The hefty pay for Goldin, 63, to stay away represents 40 percent of the $4.5 million in salary and deferred compensation Goldin would have collected under his five-year contract, a source said."

The amount paid to Goldin was reported as being between $1.8 and $2.0 million. Just to put this in perspective, at a cost (room and board) of $38,194 per year, $2 million represented a full four-year education for 13 students at Boston University. It was also equivalent to 80 NASA buyouts at $25,000 per person.

A few days later the New York Times would report, "[Goldin] was going to sweep the place clean-management, deans," one person close to the university said. "He was walking around talking about who should stay and who should go." People at NASA saw this as an eerie echo of Goldin's days at NASA a decade earlier.

Dan Goldin had just earned the equivalent of more than several decades of an average NASA employee's salary—perhaps even a career's worth. It wasn't because he had done something marvelous. Rather, he was paid not to show up for work. Nothing Goldin had done served to make anyone at NASA proud. To have their former leader behave like this—as the agency was still reeling from the aftermath of the Columbia accident and trying to bounce back was a source of embarrassment.

Once again the NASA family just wished Mr. Goldin would hurry up and fade away—once and for all.

Return To Flight (RTF)

The CAIB addressed the changes needed to the Shuttle orbiters and the foamencased tank to avoid another Columbia-like accident. Thermal Protection System (TPS) damage, caused by either ice or foam had plagued the Shuttle program on virtually every flight since 1981. The CAIB's first instructions to NASA were to eliminate critical, large ascent debris, and do 'significant risk mitigation' to enhance the safety of the shuttle and its launching stack.

To accomplish this, CAIB ordered a redesign of the shuttle's External Tank bipod assembly, since this was the site from which much of the shedding foam on launches had occurred. It was a ramp-like structure located just beneath the orbiter's nose, at the point the spacecraft attached to the tank. To replace the ramp, NASA chose to install a series of electric heaters. The heaters were to avoid the formation of ice, which the foam was initially designed to inhibit. A second CAIB requirement was met by developing a test program that would allow engineers to better understand what caused the foam shedding, and the creation of a series of alternate designs that could be employed to further reduce falling debris.

A program was started to conduct a new series of tests that would assure that various parts of the shuttle orbiter could withstand impacts from other, smaller types of debris. This was a response to a second CAIB finding that called for NASA to 'increase the orbiter's ability to sustain minor debris damage'. This was to be achieved by improving the impact resistance of the Reinforced Carbon-Carbon (RCC) as well as tiles. In the process, NASA would be able to judge how resistant the existing TPS materials actually were.

Beyond the Columbia launch and debris strikes from shedding foam that had occurred on other missions, CAIB ordered a structural integrity review of the orbiter's complete Reinforced Carbon-Carbon leading edge panel components. It would be accomplished by using new testing methods. The board also urged NASA to develop and maintain a database of all flown RCC system elements and order more spare RCC assemblies and parts.

Anticipating damage

For shuttle flights up to the space station, the CAIB required the development of 'a practical capability to inspect and effect emergency repairs to the widest possible range of damage to the TPS'. This included tiles and RCC panels. Cameras mounted on the station or held by astronauts were to be able to photograph the shuttles as they approached the station and while they remained docked to it.

For flights to other orbits, NASA was to develop an autonomous inspection and repair capability 'to cover the widest possible range of damage scenarios'. Eventually, the shuttles were to be capable of inspecting and repairing all potential TPS damage 'early in all missions'.

But the board's ultimate objective was a fully autonomous capability for repair so that if 'an ISS mission fails to achieve the correct orbit, fails to dock successfully, or is damaged during or after undocking', the vehicle could be fixed. Only once in shuttle program history, during a July 1985 Challenger launch, had a shuttle failed to make its intended orbit.

In the event of minor damage to the RCC during a flight, the CAIB suggested that NASA be able to effect reentry of the shuttle safely 'with minor leading edge structural subsystem damage'.

To further protect the shuttle heat shield before sending the vehicle into space, the board urged that NASA improve the maintenance of the launch pad structures, minimizing the use of zinc primer on the super-structures. During the investigation the board determined that flaking of zinc could corrode the leading edge RCC parts of the shuttle while it sat on the pad before launch.

At the time of the Columbia accident, NASA had used an older computer software program called CRATER to evaluate debris hits and their possible damage to the vehicles, CAIB called for development of a new computer program to more accurately develop models to predict debris damage from any materials striking the climbing shuttles during ascent. The report also called for establishment of a threshold for detected impact damage that would trigger on-orbit inspection and if needed, repair. Clearer definitions for such things as 'foreign object damage' and elimination of confusing terminology that might serve to hinder debris issues was also suggested.

Limited launch options

The CAIB called for upgrading the imaging system used at the KSC launching sites, so that a minimum of three useful views of the climbing shuttle would be had, from liftoff to at a minimum Solid Booster separation 'along any expected flight azimuth'. The systems and their readiness to be used on a given launch should be folded into the shuttle Launch Commit Criteria. Ships and other aircraft should also be used to track the vehicle during flight. From onboard, the board called for NASA to develop the capability to obtain and send back to the ground high resolution images of the vehicle's leading edge wing and the forward section of the TPS. Use of spy satellites (referred to circuitously by NASA as 'national assets') to photograph the shuttle for every mission should become a standard.

The most substantive—and daunting—recommendation by the CAIB called for NASA's safety organization to be completely restructured, and a flight schedule adopted that 'is consistent with available resources'. Astronaut training schedules and procedures should be modified as well, expanding the procedures for in-flight emergencies, and establishing an independent means for verification of launch readiness prior to flight. The board also suggested that the shuttle program office be reorganized to better integrate all of the elements of the program. These were the most significant CAIB recommendations that needed to be implemented before the next launch.

Stafford-Covey

NASA wasted little time preparing its return to flight plans. Even before the August release of the CAIB report, the agency chose a pair of former Apollo and shuttle astronauts to head up their RTF effort. Former Apollo commander Thomas P. Stafford, and former shuttle commander Richard O. Covey, were assigned to lead the RTF Task Force to both plan the RTF as well as follow how well NASA was implementing the CAIB recommendations.

Both Stafford and Covey had, "Rich backgrounds in technical engineering, safety, management, and other areas vital in expediting NASA's implementation of the Columbia Accident Investigation Board's recommendations," O'Keefe said when he announced their appointment 13 June 2003.

O'Keefe also named others to what became known as the Stafford-Covey task group. Clinton Administration Navy Secretary Richard Danzig, former Apollo 8 astronaut and General Dynamics CEO Bill Anders, former USAF Chief of Staff Ron Fogelman, former Director of Shuttle Processing Bob Sieck, former deputy to Admiral Hyman Rickover William Wegner, former director of the Congressional Budget Office Dan Crippen and the former Commander of the Space and Naval Warfare Systems, Admiral Walter Cantrell, were appointed to the group.

The RTF charter was also quickly established. The task force would perform its assessment of the RTF planning 'independent' of the agency. The former CAIB board members would be consulted. It would act as an advisory body, and would issue reports during the RTF preparations, as well as a final report. All of its members were appointed by O'Keefe, and NASA's Office of Spaceflight was to provide it with technical and staff support. NASA was also the group's funding source, and its members would be paid for their time, at the GS-15 step 10 level ($100,000 or more a year). Cost of the task force was expected to be $2 million.

They went to work immediately, with a detailed RTF plan published the month following the release of the CAIB report. NASA also created a Space Flight Leadership Council made up of the most senior shuttle program managers and co-chaired by Associate Administrator for Space Flight Bill Readdy and Dr. Michael Greenfield, Associate Deputy Administrator for Technical Programs. It would formally approve the RTF process and launch preparations. It had been hoped that the RTF mission could take place as early as spring 2004, but it quickly became apparent that it would take longer to implement the CAIB-induced changes. A launch during the period from 12 September 2004 to 10 October 2004 was now the new planning period.

NASA began a comprehensive test program aimed at giving shuttle planners a better understanding of the basic cause for the foam shedding. The agency also said it had begun to look deeper into the shuttle's design to more fully understand other sources of risk to safe flight. Concerns arising from the aging of the shuttles were to be given renewed attention, as well as new methods to assess the health of the orbiter while flying in space.

To respond to the imagery issues, a Memorandum of Understanding had been signed with the National Imagery and Mapping Agency even before the CAIB report had been released that made orbital imagery of shuttles in flight standard during each mission. New high speed and high definition cameras were to be installed in the area around the shuttle launching pads, and cameras were to be mounted on the External Tank and boosters. Procedures were to be developed that would allow real-time transmission of their shuttle views. New hand-held cameras would be carried by astronaut crews so they could better photograph the external fuel tanks as they were jettisoned.

The on-orbit inspection by the shuttles themselves would be aided by a camera attached to a new boom that would be mounted on the end of the orbiter's remote manipulator unit. The device would allow astronauts to see virtually every element of the exterior of their craft.

Pathway to full compliance

NASA also began development of procedures by which astronauts on the station would use their new cameras to image the approaching and departing shuttles. More restrictive launch requirements would be imposed on the return to flight and subsequent launches. For that first, RTF flight, this would mean a launch only in daylight conditions, and External Tank separation only in daylight.

The tile and Carbon-Carbon repair issues were tackled immediately, as NASA began to develop an onboard repair kit and other plans to store RCC panels and other TPS elements aboard the orbiter itself, and also aboard the station. It called tile repair 'feasible but technically challenging'. Should the orbiter be damaged beyond repair during an ISS mission, procedures were being developed to allow the crew to abandon the vehicles and use the space station as a 'safe haven' until another shuttle could perform a rescue.

Other CAIB recommendations about TPS databases and other planning and computer programs were also under development. A complete review of the entire flight environment of the shuttles during their missions was started, and a plan for improvements in shuttle longer-range planning and budgeting was promised. Mindful that the 107 debris had not injured anyone, an event that many considered nothing short of miraculous, NASA began a new evaluation of the potential risk posed by the shuttle's return trajectory, possibly leading to changes in the vehicle's flight paths in the future. Bill Readdy ordered the entire certification of flight process to be overhauled, plus a new review of the waivers now in place for shuttle systems, and the methods by which new waivers might be issued.

NASA addressed CAIB recommendations in two groups; those needed for the return to flight, and those needed later, during the operation of the shuttles during their remaining use assembling the station. On 15 October 2003, more than a month after the September 8th release of the original RTF plan, a first revision to the RTF plan was announced.

Tests had begun that duplicated the foam impact that occurred on the 107 flight. NASA completed the design of a new hydrogen tank/intertank flange that was to reduce the possibility of voids between the tank shell and the foam. New x-ray and other imaging systems were designed and tested.

More foam impact tests were conducted on RCC panels from the orbiter's wings. The testing didn't appear to show damage, but the photographing of the panels was ordered to verify the results.

A series of simulations of repairs to tiles were conducted in late 2003, and the 'repaired' units subjected to heating that reproduced reentry temperatures. New EVA tools were tested during KC-135 zero gravity flights.

But as 2003 ended and the new year began, it became increasingly likely that meeting the late 2004 launch schedule would not be possible and still meet the CAIB requirements. NASA had selected orbiter Discovery for the return mission, and its designated crew was training for what would basically be a test flight to the space station, proving the new inspection procedures. The Stafford-Covey task group, and the Spaceflight Leadership Council met in early February, and moved the RTF launch to no earlier than 6 March 2005.

The Spaceflight Leadership Council determined that more time was needed to evaluate the condition of the space shuttle's rudder speed-brake actuators. Inspections during the stand down had revealed that some of the actuators had been installed backwards, apparently undetected, since the rudder had first been assembled. NASA and United Space Alliance inspectors now wanted to review the actuator conditions on all of the shuttles, as well as search for corrosion to them. In addition, the agency sought to expand the area of the ET foam tests and go beyond the range of the bipod ramp.

The new boom/camera system also needed more time for installation and tests. "We've said for months that we'd be driven by milestones, not a calendar," said Bill Readdy. "When we successfully pass those milestones, that's when the space shuttle will return to safe flight."

NASA also decided to prepare a second shuttle and keep it at the ready during the RTF mission for any possible rescue. Named STS-300, NASA chose to prepare orbiter Atlantis, and have it ready for an emergency launch within 35-70 days of such a decision. The Atlantis would be modified to carry the full combined Discovery-Atlantis crews back to earth by adding new seats to Atlantis' lower deck.

The new launch criteria also played a role in the launch date change. Given the restricted lighting conditions, a launch after September 2004 would have left NASA with just three possible dates in November, none the following month, and three in January 2005.

A new RTF flight launch window would run from 6 March 2005 to early April 2005. In an update to the RTF plan released on 30 April 2004, NASA announced that it had 'closed' three of the CAIB's 15 critical RTF requirements. The NASA-NIMA memorandum covering the on-orbit photography of the shuttle had been completed, the requirement for non-destructive testing for the TPS and RCC had been met, and a requirement for 2-person inspectors during vehicle close-out had been instituted.

Wayne Hale, NASA Deputy Program Manager for the Space Shuttle, said in the five months since the last update to the RTF plan, "We've made significant progress in all the items for the return to flight." The latest tests regarding foam loss from the ET suggested the agency had made major strides in reducing the foam loss to sizes that would be well below the threshold for possible damage to the spacecraft that might require repairs in flight.

Fix it and fly it

But returning the shuttles to safe flight was still a struggle. Left unsaid, but worried over by O'Keefe, Readdy, and shuttle and USA managers was the possibility that there were other safety issues or system problems lurking undetected. NASA was determined that the second lost space shuttle not be followed by a third. To this, workers plunged in around-the-clock sessions. O'Keefe lived with the concern every day, but as 2004 wore on, the progress they made in responding to CAIB slowly strengthened his and the shuttle program's confidence that the redesigned shuttles would be as safe as they could be, and operated as safely as this system could be flown.

The public and the shuttle

Even with the return mission not yet flown, the accident and its aftermath, and the CAIB report had profoundly changed a large part of NASA. Their relationship with other federal agencies, and the public as well, was altered forever. No better example of this was the FEMA-led search for Columbia's wreckage. FEMA considered it a 'three-month effort to get the shuttle flying again', as they stated on their web site at the time. It had coordinated with NASA, the Texas Forest Service, the U.S. Forest Service, the Environmental Protection Agency, and more than 60 agencies and 270 volunteer and local groups.

Field offices for the search were opened at Barksdale Air Force Base in Louisiana and in Lufkin Texas. A satellite field office was also established in Forth Worth Texas. Huge numbers of volunteers were involved in the search, everyone from local volunteer fire departments to Boy Scout troops. The Navy conducted underwater searches in Lake Nacogdoches and the Toledo Bend Reservoir. Some 60 divers from the Navy, Coast Guard, EPA, and the Houston and Galveston police and fire departments plunged into the icy waters, looking for Columbia. More than 16,500 people braved wild hogs, snakes, vermin of every sort, and the ever-changing weather as they combed the vast expanse.

Two helicopter pilots were killed on 27 March 2003 in a crash that occurred while they were engaged in the search. At the end, they brought home 82,500 pieces of the lost shuttle. No wreckage was ever found west of Littlefield, Texas, or east of Fort Polk, Louisiana. In total, some 25,000 people participated in the search for shuttle Columbia's remains.

And the search changed them as well. When Challenger was lost, its wreckage fell into the Atlantic Ocean. Its recovery was done by a limited, specialized force. Columbia died before people's eyes as they stood on their front porches that morning to watch what they had thought would be a triumphant return from space. Its pieces fell into people's farms, communities, and lay smoldering on the very land where they lived. To them, this was their space program and it had come home to them in a very personal, if regrettable way.

Given the war that the country would soon find itself facing, the shuttle was a symbol of America in more than the usual number of ways—for it combined pride in accomplishment—and the sorrow that came with loss.

Fragile and pioneering, flawed but a marvel of both compromise and technology, it was indeed a perfect symbol of the nation.

And it would fly on.

Still, no one could ever be sure that there wasn't something, somewhere hidden and undetected in its design that could plunge NASA and the nation into another disaster, for risk was an unseen passenger aboard every shuttle flight. No one could assure Readdy and his team that such risk wasn't accompanied by an undiscovered flaw, waiting to emerge after hiding in history's shadow.

*   *   *   *   *

5

WHERE'S YOUR VISION?

It was less than two weeks after the Columbia accident when Congress held its first hearing on the topic. This was an unusual hearing—a joint hearing held by the Senate Commerce Committee and the House Science Committee's Subcommittee on Space and Aeronautics. Sen. John McCain (R-AZ) and Rep. Sherwood Boehlert (R-NY) co-chaired the hearing. Sean O'Keefe was the lone witness.

The official charter for the hearing highlighted the following questions: "Is the Gehman Board sufficiently independent of NASA? Did NASA propose and receive adequate budgets for the Shuttle program? How does the grounding of the Shuttle affect the International Space Station (ISS)? Should NASA accelerate plans for a replacement for the Shuttle? Is the Shuttle's basic design unnecessarily risky? What other programs and payloads are affected by the grounding of the Shuttle?"

None of the questions were at all surprising given what had happened.

The day before the hearing, the complaining—much of it thinly veiled partisan sniping—began. Rep. Bart Gordon (D-TN) was openly contemptuous of the way that Hal Gehman had assembled his panel and he saw NASA interference at every junction.

According to the Washington Post Gordon said, "Right now, I'm concerned that the credibility of this so-called investigation can be challenged and I think that will be a problem if we have to start all over again later because there isn't confidence this is an independent investigation."

Rep. Boehlert was rather less suspicious of the NASA/CAIB relationship. He opened his comments at the hearing on 12 February 2003 by saying, "We will also all be coordinating with the Columbia Accident Investigation Board, headed by Admiral Gehman. I've spoken to Admiral Gehman, and I am impressed with the Admiral's determination to be independent and deliberate, vowing to be swayed neither by outside pressures or artificial deadlines. And I appreciate the swiftness with which Administrator O'Keefe activated the Board."

This was not to say that Boehlert was simply going to take everything NASA said at face-value either. He added, "Despite the best of intentions, NASA has, at times, already put out misleading information because it didn't check the facts. For example, information indicating that environmental rules could have contributed to the accident has so far turned out to be entirely spurious."

That day Sen. John McCain appeared on CNN. When asked if budget cuts at NASA had anything to do with the Columbia accident McCain said that this was a "Legitimate question—one of the first things to be asked of Sean O'Keefe. We have a two phase inquiry. The first is what caused the tragedy. The second is a long overdue policy debate: What is the role of the shuttle? Will there be a follow-on? Will we be doing exploration? What is the use of the ISS? What are the budgetary considerations? These policy decisions need to be made as soon as possible."

O'Keefe was grilled for the entire morning and into the afternoon. Many of the questions were simply repeats of what someone had asked a few minutes earlier— questions which, in turn, were repeating similar lines of questioning themselves. Everyone was sad, frustrated, angry, and looking for meaning. Some sought to grandstand as well. Rep. Anthony Weiner (D-NY) who had shown little public interest in NASA made sure he got some TV time and complained, some thought in bad taste, that the research done on STS-107 could have been done if the ISS had been completed on time.

At the end of his prepared statement was a glimmer of what was now starting to form in the back of O'Keefe's mind—and the president's—with regard to where America should go in space.

"Today, February 12 is also the birthday of President Lincoln. And some of his words, spoken for a very different purpose, have come to be in my mind this past week: 'It is rather for us to be here dedicated to the great task remaining before us - that from these honored dead we take increased devotion to that cause for which they gave the last full measure of devotion.'

"We have an opportunity here and now to learn from this loss, and renew the boundless spirit of exploration present at NASA's beginning. We will do this by being accountable to the American people for our failings and, we hope, credible and compelling in pursuit of research, exploration, and inspiration for future generations."

Later that day in an attempt to yield to the clear intent of Congress, O'Keefe agreed to allow a modification of the CAIB charter which, according to a NASA press release 'removes any requirement, either real or perceived, that asks Admiral Gehman to coordinate or await approval from NASA for any dimension of the panel's investigation'. O'Keefe had agreed at the end of the day's marathon hearings, at around 1:30 p.m. to amend the charter per the wishes of Congress.

For some reason this was not good enough for Bart Gordon who repeated his call for NASA to assure the independence of the CAIB from any possible NASA influence in a press release issued just before 7:00 p.m. that evening—a press release that made no mention whatsoever of O'Keefe's agreement to modify the charter or the press release issued shortly thereafter. This would not be the last time Congress would complain about the way in which the investigation was handled; nor would it be the last time that they flogged the issue for what clearly seemed to be to score partisan points.

Later that day NASA also released the full text of the 'Agency Contingency Action Plan (CAP) for Space Flight Operations (SFO)' - the plan that Bill Readdy had been carrying with him at the time the accident occurred. Within the document were the specific steps to be taken as soon as a major incident like the Columbia accident occurred. This was just the first of an information deluge NASA would release.

Growing frustration

As the days moved ahead, a deluge of internal email, memos, transcripts, documents, and images flowed out of NASA. Sean O'Keefe and Paul Pastorek had urged that the agency issue things as soon as they had them—and they did. On one hand, the issue of openness began to fade. On the other hand, the sheer volume of the materials provided fodder for even the laziest reporter to uncover major errors on NASA's part. As Pastorek had predicted it was going to be a nasty time for all involved.

As the weeks moved on, media and congressional scrutiny seemed to uncover a major flaw in NASA's way of doing business every day. As this process went on, a pervasive, collective sense of re-evaluation began to emerge as well. The 'what' of the accident was unfolding at a feverish pace. The 'why'—as in 'why are we in space to begin with'—emerged more slowly, with the interest in the answers made more urgent with every revelation of NASA shortcomings.

As he faced one press gaggle or congressional committee after another, something started to emerge with regard to O'Keefe's take on the situation. To be certain, seven people gave their lives for something they truly believed in—something that should drive whatever the agency did in the future.

O'Keefe also showed some impatience with the way that, what would soon be called, NASA's 'culture' tended to respond to things. Some people at the agency tended to act in ways that often defied explanation to the outside world. Still only a year into the job, O'Keefe still retained much of his outsider's mindset when it came to dealing with NASA's 'culture'.

At a 10 March 2003 press event, O'Keefe was asked about something Shuttle Program Manager Ron Dittemore repeatedly said in post-accident briefings with regard to options which could have been exercised i.e. that 'nothing could be done anyway'.

O'Keefe tensed up visibly at being asked the question. He then said, "With all due respect, Ron Dittemore is not speaking for the agency in this regard. I fundamentally - absolutely - reject the proposition that there was nothing that could have been done on-orbit. The context within which he stated that has an entirely different meaning than what has been attributed to him. There isn't anybody here at this agency that, had there been a strong indication that something was wrong, wouldn't have looked at the scenarios, the contingencies to see how applicable any of these might have been. There is no doubt in my mind whatsoever."

He continued, "Given the history of this agency there is nothing—we would have spared no effort to avoid a catastrophe. Every incident you can think of will never be covered by every scenario or every simulation that you can run. But to suggest that we would have done NOTHING! at all is positively fallacious. If there would have been any indication at all that would have driven us to the point—and I am not talking about 'what if' scenarios—but rather a clear indication that says 'here is some apparent damage, or something that is causing an operational deficiency', there would have been no end to the efforts to try and figure out what could be done."

At this point O'Keefe started to get a little angry. "Apollo 13—think about the circumstances there. There was no International Space station to rendezvous with. There was no other capsule up there. There was nothing on the pad that could have gone up to retrieve them. So I am sure there could have been a lot [of people saying] at that juncture - 'Oh well, I guess there's not much we can do'. But that is not what happened. In a situation like this you have to look at what you have available. Was this [the cause of the Apollo 13 accident] in anyone's scenario or simulation? No. This was not a story that someone cooked up. This really happened. There are folks walking around alive, today, telling that story because they lived it. Not because they wrote about it on some beach weekend. This really did happen."

NASA's past was starting to bubble up—and it did not often blend well with its present incarnation. The things that the agency and its people had once done; and the way they got though hard times really impressed O'Keefe. That was not the agency O'Keefe now had to manage—one where people whined about things, where field centers acted as if they were their own mini-NASA, and how so many people seemed to want to avoid taking the responsibility for their actions. Dittemore's shrugging off of the notion that something might have been done really irked O'Keefe.

As O'Keefe sought to try and bring his agency out from the depths of this accident, the vultures were swirling about, ready to pick at everything he tried to do. At the same time, it seemed, some of his own people seemed to be all too eager to encourage this situation to continue.

Pausing to remember

In March Congress managed to agree on one thing: a memorial to the crew of Columbia. Quoted in a 26 March 2003 press release for the occasion, partisanship evaporated for a moment: "I hope that one day, young people will walk through Arlington cemetery, come upon this memorial to these heroes, these space pioneers who led the way for our country, and be inspired by their courage just as we are inspired today," said Space and Aeronautics Subcommittee Chairman Dana Rohrabacher (R-CA)."

Subcommittee Ranking Democrat Bart Gordon (D-TN) added, "It is fitting and proper that we honor the memory of the Columbia astronauts. They died serving their country and in the pursuit of knowledge. I hope that the Congress will move quickly to enact this legislation."

Sam Brownback asks "why"?

Sen. Sam Brownback (R-KS) was new to the topic of space and he had been seeking to educate himself on the topic when the Columbia accident happened. Unlike previous chairs of the Science, Technology, and Space Subcommittee of the U.S. Senate Committee on Commerce, Science, and Transportation, Brownback often used hearings as a tutorial of sorts. Some in the space industry would complain about these 'space 101' sessions for Brownback. Others would see them as a chance to get at some fundamental topics that were often glossed over by the Senate.

On 2 April 2003, Brownback chaired a hearing which began to look at where NASA should be going as it moved out from under the shadow of Columbia. Brownback had a very basic, uncomplicated interest in space at this point—one that had not yet been cluttered by what contracts might benefit in his home state, as was often a constant bias in the positions taken by others on the committee. He simply wanted to know answers to his questions and he was not afraid to ask them. The witnesses that day gave him something to think about.

Brain Chase, at that time the Executive Director of the National Space Society said "Space exploration is a worthwhile endeavor and a sound investment in the future, and it is an investment that can be made even while meeting other needs in our nation. It is important to invest in the future, and it is important, as a society, to continue opening frontiers. History teaches us that societies that have pushed their frontiers outward have prospered; those that have not have withered and faded into the history books. No society has ever gone wrong opening up the frontier, and we shouldn't stop now."

Marcia Smith, a highly respected analyst with the Congressional Research Service went to rather great lengths to characterize why America and other nations were drawn to explore space. Placing America's early space efforts into context she said, "The risks were high in those early flights. We had little experience with launching rockets into space, and with the spacecraft that protected the astronauts. Yet the nation was willing to accept those risks, and pay the cost, to ensure American preeminence. Indeed, only three weeks after Alan Shepard's flight, President Kennedy called on the nation to commit to the goal of landing a man on the Moon by the end of the decade, and the nation said, 'yes'. Although the space program has changed in many ways over the past four decades, human space flight as an indicator of technological preeminence appears to remain a strong factor."

A few days later, on 8 April 2003, Sean O'Keefe was asked at a hearing by the Subcommittee on VA-HUD Independent Agencies of the House Committee on Appropriations to characterize how the agency was adjusting its planning after the Columbia accident. He said, in his prepared statement,

"Mr. Chairman, I will tell you as I told them, I think not. A test of any long-term plan is whether it can accept the inevitable setbacks and still achieve its goals. That is my hope for our plan. Mr. Chairman, in light of the recent tragic loss of Columbia, we must recognize that all exploration entails risks. In this, the Centennial Year of Flight, I am reminded of an accident that occurred just across the river at Ft. Myer in 1908 onboard the Wright flyer. The Wright brothers were demonstrating their flying machine to the U.S. Army, and a young lieutenant was riding as an observer. The flyer crashed, and Lt. Thomas Selfridge died of head injuries, thus becoming the first fatality of powered flight. From that accident in 1908 came the use of the crash helmet. So too from Columbia we will learn and make human space flight safer."

Preoccupation

While the notion of where the agency was headed did come up, Congressional attention—and that of the media—was still focused on Columbia. The CAIB would begin to issue interim findings; each one formally capturing how NASA had erred and how it should fix things. 'What if' scenarios began to circulate as to what might have been possible if only NASA had known what the problem was. The agency was being hammered and each new revelation seemed to paint a picture of an agency ever more incapable of getting anything right.

With NASA's FY2004 budget under consideration, Congress began to question how the Orbital Space Plane was going to fit into the equation and how the Shuttle would be fixed (if it could even *be* fixed.) In May 2003 venerable spacecraft designer Max Faget told the Los Angeles Times, "The bottom line is that the shuttle is too old. It would be very difficult to make sure it is in good shape. We ought to just stop going into space until we get a good vehicle. If we aren't willing to spend the money to do that, then we should be ashamed of ourselves."

On 6 June 2003 the cause of the accident was more or less sealed. Tests involving the firing of foam from a pressurized nitrogen gun at shuttle leading edge materials showed that foam could indeed cause damage of the sort required to produce the events that led up to Columbia's destruction. The next day Columbia Commander Rick Husband's wife Evelyn told Associated Press, "I don't want to see NASA hammered over issues that are irrelevant or unfair. I just don't want there to be a witch hunt just for the sake of a national television audience ... to see NASA get pummeled."

A few days later, at a small ceremony at the Department of the Interior, Sean O'Keefe and Interior Secretary Gale Norton formally announced the naming of 'Columbia Point', a 13,980-foot mountain peak in Colorado's Sangre de Cristo Mountains, in honor of the Space Shuttle Columbia's final voyage.

On 10 July 2003 the CAIB announced that it would not be releasing its report as planned and that it would be delaying the release for more than a month until 26 August. By now everyone knew what had happened. What was yet to come was the inevitable finger pointing and harsh remedies that everyone knew were in the offing. What people also wondered was if the report would go beyond the cause of the accident to look at why America was devoting effort to the human exploration of space. Hal Gehman and other CAIB members signaled very clearly that indeed the CAIB would address that issue.

Looking for some inner motivation

Indications of how NASA was trying to motivate itself were to be found two days later when a letter written by Office of Space Flight AA Bill Readdy began to circulate within the agency. In this letter Readdy recounted his own sorrow and anger that he had experienced as a result of the accident. Speaking of the ceremony for Laurel

Clark at Arlington Memorial Cemetery he said, "Following that moving ceremony I went back to my office at NASA Headquarters. I was so very, very angry. I was angry with myself. Angry with everyone around me. Angry with the world in general. The system had failed. We had failed. I had failed. I felt the need to DO SOMETHING."

After going through a methodical listing of where the shuttle team had been, where it was, and where it needed to go, Readdy struck a confident tone. His detractors would call it arrogant. Yet everyone agreed that he was determined to move ahead with confidence. "We are forging a new leadership team at NASA whose impact will be felt for many, many years to come. First and foremost that impact will be brought to bear on Return to Flight. In so doing, we shall define not only the future of our Space Shuttle program - but also the course of NASA itself. Indeed, we aim to take this occasion to alter the course of the human exploration of space. In the process we will also redefine ourselves."

Readdy closed by saying, "And as we emerge from this trying time, we'll come back smarter, wiser, stronger and safer. We owe it to the Columbia crew, their families, the astronauts, and the nation to sail spaceward once again aboard Discovery, Atlantis, and Endeavor in the years to come. To do any less dishonors the sacrifices of the Apollo 1 crew, the Challenger crew and the Columbia crew, and for that matter, all space-faring humans. For we on Earth have a solemn compact with all who would leap above the sky for the benefit of all humankind."

Emerging interest in Congress

In late July Sen. Brownback held another hearing. He had been quietly doing his homework while observing the political winds as they blew through the post-Columbia period. Unlike many of his colleagues who were simply waiting to fire off another skeptical press release about the latest revelation from the CAIB or NASA, he was looking ahead. Although he was also openly skeptical about NASA's capabilities, he wasn't just looking at the problems. He was looking at what lay beyond the problems. Brownback put a voice to something he had sensed among his colleagues. He felt that many in Congress wanted to see the U.S. 'dominate the Earth-Moon system'. Brownback's rhetorical question was, in effect, 'What policies and programs can Congress and the Bush Administration recommend and put into place that would foster such leadership?'

Summer break consumed the Congress' attention and the media continued to detail NASA's ongoing dilemma while the CAIB labored to finish their report. Little was written about what comes next-rather everyone focused on what they thought the public wanted to know i.e. what went wrong.

At last, the report

On 26 August 2003 the CIAB report was released. The detail and dedication to detail were surprising. Not only had the CAIB documented what went wrong, they had produced what amounted to a handbook on the history and conduct of human spaceflight. While much of what was contained in the report was already publicly known, the emotional impact of its final release was hard to ignore. In addition to the detailed breakdown of what went wrong—and why—were the seeds of where America should go next.

The CAIB made some rather expansive observations in this regard. In a separate chapter titled 'Implications for the Future of Human Spaceflight' saying:

"All members of the Board agree that America's future space efforts must include human presence in Earth orbit, and eventually beyond, as outlined in the current NASA vision. Recognizing the absence of an agreed national mandate cited above, the current NASA strategic plan stresses an approach of investing in 'transformational technologies' that will enable the development of capabilities to serve as 'stepping stones' for whatever path the nation may decide it wants to pursue in space."

With regard to the space shuttle Program, the CAIB said that it had "reached an inescapable conclusion: Because of the risks inherent in the original design of the Space Shuttle, because that design was based in many aspects on now-obsolete technologies, and because the Shuttle is now an aging system but still developmental in character, it is in the nation's interest to replace the Shuttle as soon as possible as the primary means for transporting humans to and from Earth orbit. At least in the mid-term, that replacement will be some form of what NASA now characterizes as an Orbital Space Plane. The design of the system should give over-riding priority to crew safety, rather than trade safety against other performance criteria, such as low cost and reusability, or against advanced space operation capabilities other than crew transfer."

Of course what space transportation system you need and what it is to do depends on what overarching policy you have—one which determines what you do in space— and when. Absent such a clear plan or vision, such priorities are hard—if not impossible—to determine.

The CAIB noted that its investigation "has focused on the physical and organizational causes of the Columbia accident and the recommended actions required for future safe Shuttle operation. In the course of that investigation, however, two realities affecting those recommendations have become evident to the Board. One is the lack, over the past three decades, of any national mandate providing NASA a compelling mission requiring human presence in space. President John Kennedy's 1961 charge to send Americans to the moon and return them safely to Earth 'before this decade is out' linked NASA's efforts to core Cold War national interests. Since the 1970s, NASA has not been charged with carrying out a similar high priority mission that would justify the expenditure of resources on a scale equivalent to those allocated for Project Apollo. The result is the agency has found it necessary to gain the support of diverse constituencies. NASA has had to participate in the give and take of the normal political process in order to obtain the resources needed to carry out its programs. NASA has usually failed to receive budgetary support consistent with its ambitions. The result, as noted throughout Part Two of the report, is an organization straining to do too much with too little."

While not advocating any specific future destinations, or suggesting that the ISS would not be a fine place to limit activity for the foreseeable future, the CAIB did note the disconnect between overall vision and how you operate manned and unmanned access to space.

It would take some time before this general notion that clearer thinking was needed about space such that a clear national plan or vision could emerge. Moreover, whatever might emerge would require support from the White House if that policy was going to have any chance of being successful.

One last blast from Congress

Of course, the next thing that NASA could expect, once the CAIB report was released, was that the House and Senate would want to get their public chance to be heard. The first gauntlet Sean O'Keefe had to survive was a hearing on 3 September 2003 before the Senate Committee on Commerce, Science, and Transportation.

Hal Gehman went through yet another recitation of what the CAIB had found—and what it recommended. Gehman saw fit to end his prepared remarks by saying, "It is our intent that this report be the basis for an important public policy debate that needs to follow. We must establish the Nation's vision for human space flight, and determine how willing we are to resource that vision. From these decisions will flow the debate on how urgent it is to replace the Shuttle and what the balance should be between robotic and human space flight, as well as many other pressing questions on the future of human space flight. Let the debate begin."

Sean O'Keefe focused on comments he would repeat many times in the coming weeks in response to Gehman's findings and recommendations with how NASA was going to respond; and in some cases, had already begun to respond. With regards to where the nation should go next in space, O'Keefe said, "I believe it is important that all 13 CAIB members arrived at and agreed to the final conclusion of their report: 'The United States should continue with a Human Space Flight Program consistent with the resolve voiced by President George W. Bush on February 1, 2003: Mankind is led into darkness beyond our world by the inspiration of discovery and the longing to understand. Our journey into space will go on.'"

Sen. McCain would make some general comments about the report and then say: "The report reminds us that we are still in the developmental stages of space transportation, and that space is an unforgiving environment which challenges our technical expertise. It also raises a number of important issues that will have to be considered as we plan for the future of the space program. Most importantly, we will have to figure out where we want the space program to go, and what we expect to get out of it."

Sen. Brownback took a giant stride out ahead of everyone—including O'Keefe by saying, "I am committed to authoring a reauthorization bill for NASA during this Congress, and will use this report to provide some of the guidelines for that bill. I am also pleased to see that the Board recognizes the importance of a vision for a future in manned space exploration by Americans, and believe it is time to look at creating a Presidential Commission on the Future of Space Exploration to establish this common vision. I have held several subcommittee hearings over the last few months with not only NASA and other Federal Agencies but also with private sector companies and entrepreneurs in an effort to ascertain what America's vision for future space exploration should be. In all of these hearings, one thing stood clear, Americans continue to support human space flight and exploration. NASA and space exploration remain a symbol of national pride."

Brownback was clearly interested in attending to NASA's problems, but he was also equally, perhaps more than equally, interested in what lies ahead. Alas, some present at the hearing were far less optimistic.

Sen. Ernest Hollings (D-SC) said, "My concern is that we have been here before and that NASA has a terrible track record. I'm not sure that NASA can reform itself. We in the Congress may need to help them, whether it's through new institutions or by changing the program's responsibilities."

… then the House

The next day the venue was on the other side of Capitol Hill. On 4 September 2003 the House Science Committee held a companion hearing to the one the Senate had held the day before. Initially the staff of the committee had been saying that there would be at least one hearing a week through September into early October dealing with the CAIB report and NASA due to intense interest in the topic by the Committee members. This prediction never materialized.

Sean O'Keefe was not a witness, but several members of the CAIB were. According to the Hearing charter for this day 'The report sets the stage for a thorough public policy debate regarding the future of human space flight, the prospects for a shuttle replacement, the appropriate balance of human and robotic missions, future priorities in space exploration, and the level of resources that should be allocated for such activities. The Committee's findings will, among other things, form the basis of a NASA reauthorization bill next year.'

Committee chair Boehlert ended his opening statement by saying, "finally, we need to better define NASA's overarching human space flight vision-something that has been lacking for more than a generation. That won't be easy, and it can only be done after hearings that will enable us to make a clear-eyed appraisal of the costs, benefits and risks of different options."

Rep. Ralph Hall, still a Democrat at that point, echoed Boehlert's closing comments very closely, saying, "Finally, we need to set some concrete goals for human exploration beyond the Space Station. Establishment of human exploration goals would ensure that we make the appropriate investments in our space program, would revitalize the NASA workforce, and would serve as a source of inspiration for both the NASA workforce and the American public."

The hearing then proceeded as had others with the witnesses, led off by Hal Gehman, going into the process of generating the CAIB report. As it came to a close it became clear that NASA had a rather Herculean task ahead of itself-just to get the shuttles flying again. Already, fears of unrealistic schedule pressure, inadequate cost forecasts, and concern for safety were raised. The only way NASA was going to cast off this doubt was to do penance, and, with due diligence, fix the shuttle fleet and get it flying again. Only then would the cloud of doubt begin to lift.

Or so some people thought. In reality there was a less visible, yet very palpable feeling that there had to be a compelling reason why we did things in space. If nothing else to put the sacrifice of Columbia's crew, and of previous crews, into perspective.

The White House was already tackling that issue. Indeed, it had been doing so for months.

What are they up to?

As Summer drew to a close and Fall approached, the Columbia Accident Investigation Board's hearings wound down; much quicker than many had expected. Now that Columbia accident questions had been satiated—at least for the time being—there was one remaining complaint from members of Congress. What is the purpose of the space program. What was it for? And where should it go?

The politicians' complaints had grown louder over the summer about the so-called NASA 'vision', or lack of one. The drum-beat for a 'vision' and/or a 'goal' would soon become relentless. Months later, in January 2004, after their desires were met with the announcement of the vision by President Bush, Sean O'Keefe would chide one member about their earlier complaints, saying, "Be careful what you ask for. You just might get it."

But back in the last warm days of summer, members of Congress who had been hearing rumors about the interagency space policy review within the White House, were now complaining about it. They weren't complaining that it was being done, rather, that the Bush Administration was making space policy in secret; in a way, they claimed, that Bush had done in formulating many other policies.

And they insisted on being heard.

Previous visions

The only detailed House space-related legislation to emerge in recent years had been a bill titled 'The Space Exploration Act' which was introduced in the 107th Congress and again in the 108th Congress by Rep. Nick Lampson (D-TX). While well intentioned, the bill was a smorgasbord of projects likely to cost hundreds of billions of dollars tagged against ill-conceived milestones.

The bill was often referred derisively as the 'Clear Lake Full Employment Act' (referring to a suburb inside Lampson's district outside Johnson Space Center in Houston). Only Democrats signed on as co-sponsors and no interest was shown by the House Science Committee to mark it up. As it had in the previous Congress, Lampson's bill languished in the shadows, once introduced.

Thus there was a vacuum of serious space policy legislation in the House. There was even less creative thinking in the U.S. Senate. And whatever the process was going to be, Congress wanted to be involved. Members were less than pleased to learn that the White House had been conducting a process for several months, without any congressional input.

During a hearing by the House Science Committee's Subcommittee on Space and Aeronautics in September 2003, a small group of Congressmen led by Reps. Ralph Hall (then D-TX) and Bart Gordon (D-TN) grilled Sean O'Keefe as to who was involved in this stealthy White House space policy process. In a manner more befitting a prosecutor, and somewhat reminiscent of the Watergate hearings ('what did you know and when did you know it?'), the committee members went through a list of names asking O'Keefe whether or not they were involved, to which O'Keefe replied either "yes" or "no."

A series of press releases and verbatim transcripts of the exchange followed later in the day reiterating the frustration of some members of the committee. Clearly they were out of the loop—and they did not like it. Meanwhile, the White House, in the form of its NASA Administrator kept its own counsel on this matter.

A few days later, on 16 September 2003, prompted by this exchange, President Bush acknowledged the planning effort but did not go in to the specifics of its activities other than to say, "We've got an interagency study going on now that will enlighten us as to the best recommendations necessary for NASA to proceed in a way that is a good use of taxpayer dollars."

A series of congressional hearings would ensue, in both the House and Senate where Sean O'Keefe, CAIB chair Hal Gehman and a number of individuals representing themselves, their universities, agencies, or organizations, testified. Everyone had a different idea of where to go in space, and in many cases, the same number of places where not to go.

The only consistent theme to emerge was that America's space agency needed a clearly defined vision to guide it; and a plan to make that happen. The prevailing consensus was simply what everyone hoped for, not what they expected to achieve.

Send in the Veep

On 20 October 2003, a motorcade swept beneath the Capitol carrying Vice President Dick Cheney and three White House aides. Also along was Sean O'Keefe. They climbed out of the limos on the Senate side and convened in Majority Leader Bill Frist's offices. Cheney had come to hear the leading Senators with interests and responsibilities in space on their views about the future of the space program. He was not about to tell them how far along the process had come. All he told them was the Bush Administration was open to their ideas.

Indeed, the White House activity had already eliminated all of the radical ideas. Mars, as a singular destination was now out—but preparation to develop the ability such that a decision could be made to go there at a later day was not. The Bush team was thinking incrementalism and a longer view; not stunts or so called 'flags and footprints' missions.

Five Senators had been summoned for the meeting; three Democrats, Fritz Hollings (D-SC), John Breaux of (D-LA), and Bill Nelson of (D-FL). The two Republicans; John McCain of (R-AZ) and Sam Brownback (R-KS) were chair of the full Senate Commerce and Science Committee and chair of its space subcommittee. Cheney opened by thanking the members for coming and saying his purpose for gathering them together was to hear their thoughts about the future of space.

Each member was allowed to bring one staffer to the meeting. Hollings was the most critical and loud. He told Cheney that O'Keefe was, in essence, a failure. "They put in charge of safety a guy who did not listen," Hollings railed. "The space program over there is collapsing." After a moment he added, "They have this culture problem nobody can get at."

Nelson said that NASA had been starved for funds for too long, and it was also time that the top leadership of the nation get involved. "You have to lead NASA from the top down," he said. "It was time OMB freed up more money," such that NASA could better do the job the country expected of it. "The vice president needs to get OMB to give NASA money to cover whatever the vision becomes," he told Cheney. "Build today for tomorrow's vision."

Breaux also worried about NASA's lack of adequate funding. "There should be an alternate to the shuttle available now," Breaux said, but instead NASA "just keeps patching up the shuttles. You need to get the resources to do it right!" Breaux told Cheney, looking at O'Keefe.

Brownback laid out four programs, from shuttle to station, where NASA had failed. "You are doing good with robotics, but not so good with manned," he told Cheney. He also urged that a new national space vision be developed. "The ISS isn't giving us a vision," Brownback said. "Get your agencies together and let's discuss going to the Moon or Mars, or whatever."

McCain expressed frustration with the costs of returning the shuttle to flight. 'We've got to know what the real costs are, Sean," McCain told the NASA chief. Little known outside the Congress, the fiery man from Arizona and the Irishman in charge of NASA were friends. McCain respected O'Keefe's management skills, but he also was deeply suspicious of NASA itself. Still, McCain said there was some reason for optimism. "If that Mars probe finds ice, it will be exciting for the country," McCain said. "What is the future of the shuttle?," he asked O'Keefe. He hailed Gehman's report. "The CAIB report was excellent," he said. But he, too, expressed concern about a lack of a cohesive, all encompassing vision for space. "Let's prioritize and then fund appropriately," he suggested. NASA wasn't budgeting appropriately and hadn't for years. Otherwise, "You're heading for a train wreck."

All members expressed doubt about the underlying rationale for developing the Orbital Space Plane (OSP) and how NASA would be able to both operate it and the shuttle fleet at the same time.

"NASA still isn't listening," Hollings said, looking directly at O'Keefe. The others were silent.

Cheney said that there was always a debate concerning the proper mix of robotics and humans in space. Nelson shot back, "You've got to do both!" Hollings again addressed O'Keefe. "What is the situation with the OSP?" he asked. "What are you doing?"

O'Keefe spoke up, saying that the OSP was needed to increase access to the station. Or at least something that performed that function. "We will finish the station. All of the elements are ready when the shuttle starts flying again. We need your help in the '04 budget."

Hollings spoke again criticizing NASA under O'Keefe. "You didn't take those pictures," he said, referring to the decision not to photograph the Columbia in space during its mission.

Cheney decided to cut off the meeting since he had a similar session planned on the other side of Capitol Hill at 3:05 p.m. with House space leaders. "Thank-you for your input," he told them. "We'll keep you advised as we work through this." Of course, that was exactly what the Bush administration was wary of doing. "If we had told them what was going on," one source said later, "they would have torn this thing apart by now."

The meeting wasn't the end of congressional concerns with the space program. Policy or no policy, many felt that the administration had gone about the whole issue the wrong way. The steady drum-beat from the hearings and from the CAIB report deepened their collective frustration with NASA. It wasn't so much as they didn't trust O'Keefe. Many liked him personally. Many others, like Rep. Ralph Hall found that Bush made a point whenever he met with members and space was mentioned, that Bush would say Sean O'Keefe was his man at NASA. So they had to swallow their feelings about O'Keefe to go along with the White House.

But from time to time, both at NASA and during his brief stint at OMB, O'Keefe had occasionally had run-ins with members where he said, basically, "If you don't like what I'm doing, well too bad." When the Columbia accident happened, many thought initially that they now had their opportunity to rip into O'Keefe's secretive methods. And some members, like Rep. Sheila Jackson-Lee (D-TX), did just that, in front of the cameras during the early hearings. At one point, she waved about a small model of the shuttle, trying to show her mastery of the facts then emerging, about the foam and its potential to have caused the disaster. But none of it seemed to stick to O'Keefe. He was the White House's man, both Bush and Cheney stood with him. Try as they might, several members found it nearly impossible to shake the administration's faith in the Irishman. And that didn't sit well with many of the politicians.

Senator Hollings was so frustrated with the so-called 'transformation' at NASA that he introduced a bill 'National Space Commission Act of 2003' a few weeks later. The bill would go nowhere and would eventually be outright eclipsed in January 2004 as the president announced his space policy and along with it the creation of the Aldridge Commission. Indeed, in June 2004, Aldridge's final report would itself recommend the creation of a permanent space advisory body.

Frustrations and expectations mount

A few weeks later, on 18 November 2003, House Science Committee Chair Rep. Sherwood Boehlert (R-NY), also frustrated by the process, said, "If we're going to take on ambitious new human missions—and I think we should—then we can't indefinitely perpetuate the existing elements of the human space flight program. We need a date certain to stop flying the Space Shuttle and to decommission the International Space Station. Obviously, both will remain in use until the end of this decade and probably beyond. But while they're in use, we need to ensure that they are, to the greatest extent possible, contributing to our longer-range missions."

At the same time, the Deputies Committee, meeting at the White House, had drawn towards a recommendation for the president: return American astronauts to the Moon, and utilize their lunar missions to develop new technologies to go elsewhere. The moon was to serve as the initial focal point for all of NASA's human spaceflight objectives beyond Earth orbit.

But when Bush received a preliminary review of the likely recommendation from

O'Keefe, he wasn't satisfied. Bush had his own vision for NASA all right. But it was all about getting out there and exploring. He made it clear to O'Keefe and to Steve Hadley that he wanted a broader exploration theme, a broader objective that NASA should follow. "Get back to the drawing board," he suggested, and think in those wider terms.

*   *   *   *   *

6

IN SEARCH OF A MISSION

Mission? What mission?

What purpose does NASA serve? And why did anyone really care? These questions had gone unanswered for decades. Space exploration, and the space policy that served as a roadmap to its execution, had been a sort of handmaiden of national policy, shaped as much by benign neglect as by any sense of urgency on the part of either recent presidents or Congresses. John Kennedy had given NASA its biggest mission and biggest budget boosts, but the high priority that NASA had in the Apollo era could hardly be sustained. Its decline actually started under Lyndon Johnson, a spacer president if there ever was one, but also a president squeezed by the costs of the Vietnam War and the 'Great Society' poverty program. Both initiatives soaked up Johnson's budget dollars, to the point a tax surcharge was imposed on the public late in Johnson's term. By the end of 1968, as he prepared to leave office, he would tell his NASA Administrator James Webb, also a personal and political friend, that his cuts to NASA were forced upon him, and that he had always hoped that he could add money to the space budget the next year.

Richard Nixon was the political father of the Space Shuttle, but it was a reluctant parentage that Nixon came to agree to in 1972. He did so mainly for the jobs boost that it would sustain in vital electoral-vote rich states such as California and Texas. Still, when he had the chance to kill the shuttle as it was being born, he supported it instead. It was also the least expensive new human space program that he could initiate. Neither Gerald Ford nor Jimmy Carter had much use for NASA or the space program either, appointing politically weak administrators. But Carter had saved the shuttle from cancellation, despite the desires of his Vice President Walter Mondale, a longtime critic of manned spaceflight in general and the Space Shuttle in particular.

Under Carter's watch the shuttle's cost soared out of control, but the 39th president had resisted the temptation to terminate the program, mainly because of the national security payloads that the shuttle was to launch, all considered vital to arms control, a Carter emphasis.

New space initiatives

It would be Ronald Reagan that gave NASA its largest new program start since Kennedy and Nixon. But the permanent space station that he proposed languished, subjected to arguments over its priority and importance both within the White House staff and the Congress. George Bush, the 41st president, would actually propose a massive package of human space goals, embraced by a presidential Policy that actually called for sending Americans to Mars. But the proposed program, called the Space Exploration Initiative, failed to gather any major political or public support, and eventually died because of it.

The Clinton administration also gave space activities a low priority, mainly looking at the space station, which it initially sought to kill, as a tool of foreign policy. Dan Goldin, whatever else his faults at the end of his term as NASA chief, can be credited for saving the station in Clinton's early first years, by ushering in the Russians as partners. But that last minute save not withstanding, the station program would be vastly reduced in scope and capability during the Clinton years.

By the time George W. Bush took office, however, there were some largely overlooked staffers in the White House that formed the corporate memory for space policy, and whose work served as the foundation for much of what was to come. While Bush had not spoken about space during the campaign of 2000 except in the context of military space and missile defense, he did support an overall space policy review conducted under the direction of Vice President Dick Cheney and National Security Advisor Dr. Condoleeza Rice. A series of space policy updates were to be forthcoming from the Bush White House, the first of which addressed national policy on remote sensing systems. Following its release in 2002, a second policy review was underway on space transportation, and was in its final stages when the Columbia accident occurred, forcing a review of the review.

A critical—if invisible—role

Two staffers played important if not crucial roles in this space policy activity. Gil Klinger, of the National Security Council (NSC), and Brett Alexander of the Office of Science and Technology Policy (OSTP), were largely responsible for these early Bush space efforts. Klinger and Alexander were little known outside the Washington space community, but their influence on shaping policy was substantial. And Klinger in particular would come to play an even larger role in helping to advance the space policy-making process as the fall of 2003 drew to a close. Older, and with more experience and political skill than some of the other, younger White House staffers with a space interest, Klinger would help steer the early meetings that would eventually yield George W. Bush's space agenda. But it would be a role that would be largely overshadowed by NASA and Sean O'Keefe. But even some at NASA were envious of their White House roles in space matters.

After Columbia

While Sean O'Keefe would soon begin to forge the idea of extending NASA's responsibilities with an infusion of new presidential space support, the path that would lead to a new space policy being accepted by President Bush actually began without either his or NASA's participation. Soon after the Columbia accident, a series of meetings began in the Executive Office of the President (EOP) that sought to address the future of the human spaceflight program. These meetings were at the lowest possible level-the staff level. People were basically getting together, on their own initiative, to kick around ideas. The Columbia accident had happened, of course, and the future of space exploration appeared uncertain. They wanted to see if some clarity of purpose for space could be defined, and if they themselves could come to an agreement about it.

True believers

The largely young staffers came from a wide variety of EOP agencies. The Office of Management and Budget (OMB), NSC , OSTP, and other agencies were represented. Who were these White House employees? None of the participants interviewed for this book wanted to be identified. But what they all had in common was their love of space. They really believed in U.S. space exploration. Some had impressive technical educations; these were engineers who could talk the technical side of space flight as well as any geeks. Ironically, they weren't in engineering jobs in the White House; they were there to work in space, space planning and space policy. Anything to be 'doing' space stuff.

Splintering space

Others had policy or administrative backgrounds, but had spent years poring over every tome and paper about space that could be found. They lived and breathed the stuff. When they went on trips, it was to space museums or launches. When they got together for lunch or on their own time, it was to talk space, or attend lectures about space. And, unlike many in Washington, they actually knew what they were talking about.

Together, they now sat down in the aftermath of the shuttle disaster to think; what was America going to do now? Was there something, they thought, that the administration could do, something that could possibly be elevated to the presidential level that could the basis of a presidential announcement, or an acknowledgment by the president about the importance of space? As they continued their talks—unstructured and without any formal agenda—their meetings took on a name. They began to be called the 'Splinter Group'.

Several attendees thought that the upcoming Centennial of Flight anniversary of 17 December 2003 might serve as the perfect venue for Bush to say something about supporting the human spaceflight program. But the staffers had no real process going, not hoping or expecting to create any new policy. And they all agreed on one other thing: they had to keep things quiet, under the radar, so as not to arouse NASA in any way. As such they kept it to just a discussion.

Each time they got together, everyone in the room felt strongly that something needed to be done about the space program. Nobody wanted to entertain the thought of ending the human exploration of space because of Columbia. These 'roundtables' went on through April of 2003. As Sean O'Keefe continued to brief President Bush on the progress in the Columbia investigation, the ideas generated by the Splinter Group began to accumulate. Some brought or referred to white papers on various space topics written years before.

Then there was a subtle shift in the Splinter meetings. They became just a bit more formal. In addition to the original members, invitations were extended to other EOP officials and staff. These included staff of Vice President Cheney, some from other elements of the Domestic Policy Council, other cabinet representatives. The attendees started to look across the material that had been generated since the winter, and ask hard questions. Where should human spaceflight go? Should America be doing it at all, should the country be going all out? Many flat-out said the country should finally get going out beyond low Earth orbit, as it had during Apollo.

Together, the ideas were drafted on paper into a 'strawman' draft policy document. But they weren't yet sure if there was anything there that could eventually rise to the presidential level. They were asking each other the same questions that their predecessors had raised decades ago. "What was the justification and rationale for human spaceflight." They went around and around, meeting after meeting. What emerged looked a lot like policy prescriptions that had emerged in earlier eras: yes, human spaceflight was a good thing to be doing. Yes, it strengthened the country. Was good for national security, created jobs, etc.

Graybeard speaks

While all of this was going on, there was yet another influence being directed towards the president and the idea of a new space policy. An old family friend of Bush's father had contacted the White House. He was granted the courtesy of seeing the President in late spring 2003. Instead of saying 'hello', the old academic pressed Bush to reinvigorate the space program. Bush seemed unfazed by the request. He suggested his father's friend, who had worked closely with the Strategic Defense Initiative back in the Reagan days, write up a paper and submit it. The academic did, and wound uphaving one meeting at which he promoted a novel concept: use manned lunar exploration as the centerpiece of the civil space program.

Throughout the summer and fall he continued to press for the Moon. But he wondered sometimes if anyone was really listening.

The CAIB investigations were underway, too and the staffers meeting in the Splinter Group talked about what they were hearing from watching those activities as well. They all assumed that the Space Shuttle would be, and could be, fixed and would fly again. But there was also a consensus that was forming up that looked at the shuttle as holding the space program back, in a sense. "Should we be putting our resources into something else?" some were asking. And if so, what should that something else be? Summer was now upon Washington, and the Splinter attendees were picking up indications that NASA was interested in an expanded mission. It wasn't clear how that word filtered down to them, for none knew about O'Keefe and Bush's talks during the spring and early summer.

NASA, for its part, apparently was interested in going on its own way, 'it's own trajectory'. Sean O'Keefe, and NASA Comptroller Steve Isakowitz had their own ideas. But what they wanted appeared, as far as some in the White House heard, to be everything imaginable. O'Keefe wanted the new Orbital Space Plane proposal of his accelerated. He needed to pay for the Columbia investigation, keeping the shuttles flying at least until the space station was completed, and maybe longer. New cargo vehicle studies. Advanced rocket engine and launch vehicle technology. But what they didn't hear was O'Keefe attempting to sell a specific exploration proposal. He didn't want the Moon. He didn't seem to want Mars. All he seemed to want was a pot of new money for NASA and some new mission.

With the space agency growing more active in the search for a new vision and mission for itself, the Splinterees decided it was time to let NASA into their deliberations, if they were ever to amount to anything. "It was clear it was going to be hard to do this without them."

Rump to go

Thus the 'Rump Group' was born. A more structured version of the Splinter meetings, and with a more formal agenda and somewhat more senior staff in attendance. NASA was now represented, by three designees sent over by Sean

O'Keefe and sworn to secrecy. Steve Isakowitz, Office of Biological and Physical Research Associate Administrator Mary Kicza, and NASA Chief of Staff John Schumacher joined in. Not every meeting was attended by all three. Other senior staff from the same suite of EOP agencies joined the discussion.

The options on the table ranged from everything NASA was currently doing, as well as more. Now, as summer's end approached, the Rump Group started, for the first time, to attach cost estimates to the scenarios that they were tinkering with. Oh, you want to do that? When? Let's run the numbers and see what will it cost. Immediately it was clear that cost would rapidly constrain some ideas. So they asked that question, too: are we going to be constrained by cost in what we do? How can we work this?

For NASA's part, the answer was obvious: give us more money. A lot more money, and we can go really far out there. The ideas quickly got ranked by complexity and cost. And it was now time to take these studies, ideas, and projections and see if the White House of George W. Bush could fold them—or parts of the Rump scenarios—into a policy construct.

The Deputies speak

The policy construction process would follow a traditional form. A 'Deputies' Committee was formed up, administered by the President's Domestic Policy Council and the National Security Council. NSC, as usual, chaired this policy effort. Virtually every policy proposed for implementation had a national security element these days, one attendee thought, so having the NSC chair wasn't unusual at all.

The attendees were usually the deputy secretaries of the cabinet-level departments, although sometimes O'Keefe was present and sometimes Schumacher. The level of interest by these officials quickly became apparent, for not all of the deputies attended these meetings, or even sent representatives. One, surprising deputy that quickly grasped the importance of a new space exploration agenda, was Colin Powell's Deputy Secretary of State, Richard Armitage. To the surprise of many, Armitage not only came out for NASA, he came out loudly and never missed any of the meetings. Of course, some should not have been surprised, because Armitage was a long time friend and colleague of Sean O'Keefe. So was another attendee, Deputy National Security Advisor Steve Hadley. Hadley had often attempted to explain to the attendees Bush's active view of the reason for space exploration-exploration. It just made sense to the president to look at it that way, he said.

More Rump

Meanwhile, O'Keefe was pushing for a powerful budget boost to pay for any new initiative. Most attendees of the Deputies meetings thought he would need a big new price tag to pay for something. O'Keefe would soon collide with the White House. It had set a clear limit for budget growth that domestic agencies could follow. But NASA had two sets of numbers it was following. One was an 'in-guide' budget, and the other was an 'out of guide' budget, which basically included everything Sean O'Keefe could think of for his agency.

This didn't sit well with some at the White House, so Hadley decided to reconstitute the Rump Group—a 'sort of Rump Group II'. It was made up, once again, of just staff, but this time they had specific instructions. They were to look at some options that could be afforded that would give President Bush a new space plan without needing a massive budget increase, the kind of which O'Keefe was trying to get. First the Rump team developed detailed 'vision' options. Then they were required to run the budget numbers out through completion of the particular 'vision'.

The White House was sending O'Keefe a message: keep all of this 'budget neutral'- that is, no new money. At the time, O'Keefe had been told his agency's funding for the period between FY 2005 and FY 2009 was flat, and after that a moderate rate of growth. O'Keefe was frustrated. How could NASA get a new space vision with no money to pay for it? Everything was on the table, continue the shuttle and station, OSP, Space Launch Initiative, New Generation launch Technology program 'you name it.' NASA wanted it all, and a new exploration project of undetermined origin, too.

O'Keefe didn't say the Moon, and didn't say Mars. He wanted something on that scale, however. But he was, according to some on the Rump II staff, in a fighting mood. "This will make us (NASA) look ridiculous," he was quoted as saying in one meeting. "If we have to stay within these guidelines, there is no way our vision will get picked."

He was right about that. Oddly enough, when each of the Rump II attendees sent back their own 'vision' for NASA, all fit in the existing NASA budget plan. In their words, they were all budget neutral—except the budget proposed by NASA.

The Deputies took this data, and came back with a compromise approach. Do an analysis, they told the Rump II team, using a five percent increase to NASA and a five percent decrease. NASA came in boldly asking for a one third increase. The fivepercent decrease was called a disaster. "Why won't we just go out of business?" one Deputy is alleged to have said. Some on the president's staff had enough of NASA's whining. They were unreasonable about this budget issue, some thought. And Sean O'Keefe thought that time was running out. So he began to lobby, hard. Halloween was fast approaching. He called presidential Science Advisor John

Marburger for support. He called in other agencies. Some thought he called Bush. In the end, it began to be clear to O'Keefe that he wasn't getting any new money. It looked bleak for NASA. He called in every chit he was owed. The bean-counting administrator whom many had dismissed as being uninterested in space was now fighting-and fighting hard-for space exploration after all. He went to OMB, and the agency ran scenario after scenario.

Small change

O'Keefe, who was the previous Deputy Director of OMB, pleaded his case to his former boss, Mitch Daniels. NASA was fighting for its very existence, he said. They had to get some budget relief to achieve the goals O'Keefe knew Bush would support. He begged. He pleaded. He argued. And, finally, he won. When some staffers heard that he had wrangled new money out of OMB, many were astonished. The previous orders from Bush had been clear: no domestic agency would be getting any budget increase starting in 2005. No agency that is, except NASA. But it was a very small victory. NASA could get some new money after all, but not a lot; maybe $1 billion across the next five years. After that, nothing but flat funding, perhaps indexed for inflation.

Paying for the Moon

But how would that work? Answer: only if NASA reprogrammed some of its own, previously planned budget money. Doing this would mean that wholesale cuts to existing NASA programs would be needed to free up funding for new exploration initiatives. And that initiative appeared, as Halloween came to the Capitol, to be the Moon. A strong consensus was forming up in the Deputies group that the return to the Moon by American astronauts should be the new national space goal.

By November 2003, it was conceived that NASA would pay for it, barely, by basically cannibalizing itself. But how? For starters, it would save billions by retiring the Space Shuttle fleet as soon as the International Space Station could be finished. Billions more could be freed by killing the OSP, Space Launch Initiative, and various launch technology programs. After all, if NASA was going to the Moon, why did they need to spend billions on a new launch vehicle for the space station? And since the purpose of the space agency would become manned lunar flights, why did NASA need to be the main user of that space station? More billions saved, by canceling or drastically reducing its research into microgravity science. O'Keefe would have to impose a vast set of changes upon his agency. But by Thanksgiving, it appeared that he could pull it off after all.

Explore, okay?

The Deputies Committee looked at that plan, and thought it could work. OMB was now on board. There was just one problem. When a preliminary version of the space policy reached Bush, he wasn't happy. It wasn't about the Moon, he said. It was about exploring. The solar system. A broader set of exploration objectives. Maybe Mars in the future. Go back and make sure that's what the policy would say, he said.

So a shift was made that used the lunar surface as a sort of test bed. The technologies developed there could sustain a Moon base or help plan a Mars mission. The Deputies Committee drew up its final recommendations. The choice of 17 December 2003 as an announcement date could not be met, thanks to the budget battle delays that had raged. But the space plan was coming together and the Deputies Committee was about to go out of business, having produced a space policy recommendation that had unanimous support from the federal departments that had participated.

Quietly, but sometimes forcefully, Klinger worked to make sure the policy got structured and put into final form, with all the right players signing off. One of the staff that watched him work that Fall called him 'a powerhouse'.

Scaling back NASA

That other powerhouse at the opposite end of Washington, Sean O'Keefe, had been rearranging the budget numbers long before the chance for a new vision became possible. When he had walked in the door at NASA headquarters in 2001, he had been handed a $4 billion cost overrun on the space station program by outgoing administrator Daniel Goldin. The previous year, NASA had estimated that the remaining cost to finish the station was eight billion, paid out from FY 2002 to 2006, the date it was expected to be completed.

In January 2001 the additional $4.02 billion was added to the cost of ISS. It grew to $4.8 billion by June of that year. A task force discovered another $366 million would be needed between August and October. That put the space station, NASA's signature human spaceflight project, at a total projected cost of $30 billion. According to the Congressional Research Service, that was 72 percent above the estimate made in 1993.

It was also $5 billion higher than the limit set by Congress.

Live within your means

But NASA would get no new funds from the Bush administration to cover the overage. Legislative limits would be imposed. The Administration would cancel station elements to pay for the deficit. It canceled the station's planned propulsion module, deferred the Habitation Module, Node 3, and the planned crew return vehicle, the X-38. Basically, that ended the construction of the station at its 'core complete' stage. The price tag for that stage of the station was estimated at $8.3 billion from FY 2002 to FY 2006. But even the $8.3 billion would be inadequate.

So, under O'Keefe's orders, in December 2002, $706 million was shifted into the station budget from planned FY 2004 to FY 2007 funding, another $600 million to do core complete, and $46 million in FY 2004 to preserve long lead time construction plans. The Orbital Space Plane that O'Keefe proposed to serve as a crew return vehicle for the station would cost tens of billions more, but its costs would not be charged to the station's accounts. The source for the funding was an O'Keefe raid on the Space Launch Initiative, something that Bush had indicated would not be done, prior to O'Keefe taking office at NASA.

For FY 2005, to begin his new exploration plan, O'Keefe was granted by OMB a budget request of $16.2 billion, a rise of $866 million over the previous year. But the bulk of the funds would go to pay for the shuttle's return to flight. It would be future years, beginning in FY 2006, where the small rise in funding that he had fought for would begin.

Long after the budget battles had ended, and the policy had been rolled out, the Congressional Research Service had pegged the cost of the return to the Moon at $64 billion by 2020. The cost of going any further out into space, such as Mars or beyond, was an unknown.

The space staffers had long since returned to their duties at their respective agencies as 2004 headed to summer. It had been a year since the Splinter Group discussions and the Rump Group debates. A year since what they had started so modestly had turned into a comprehensive space policy, much to their surprise. They had learned much about space policy, and space politics too. The process by which that policy had been created didn't follow what some had been taught in school. The actual process by which policy had been made was untidy at the beginning, unformed, and wound up being pushed forward by an unusual and unlikely cast of characters, at a pace that surprised them all.

In other words, more like real life.

*   *   *   *   *

7

FALSE STARTS

A new old thing

The new plans and projects announced by President Bush were not all that new. The Crew ExplorationVehicle (CEV) announced on 14 January 2004 was the direct descendant of 50 years of thinking about spaceflight. Given all the thought that has gone into the notion of flying people and things into space, NASA has multiple architectures to do these things—just like Baskin & Robbins has ice cream flavors.

The fact that Bush endorsed such a plan is what was interesting. No president had done so in half a generation; and the last large space vision (endorsed by Bush's father) evaporated soon after the price tag became known. Indeed, no president had ever endorsed such a broad space vision—and then seen the master plan actually go on to be implemented in a generation and a half—not since John Kennedy had sent America to the Moon.

While NASA hadn't been tasked to send people somewhere in particular for a while (other than the space station), it had been tasked to do a number of large efforts to enhance its launch infrastructure so as to make such decisions easier to implement. Almost without exception such efforts either fizzled or failed.

Prior to the announcement of George Bush's new space policy, three attempts in the 1990s and early 21st century had been made to reshape America's human space flight capability: specifically, to replace the space shuttle fleet and the way America launched things into space. One effort failed miserably, the X-33. The next effort, the Space Launch Initiative (SLI), was born out of the ashes of X-33 and encompassed a haphazard smorgasbord of technologies, systems, and vehicles which slowly transformed into something else when it lost its way. The last effort, the Orbital Space Plane (OSP) initiated under Sean O'Keefe, was the development of a small human transport. The OSP project ended before it even began.

O'Keefe sought to develop the OSP in the context of a revamped ISTP (Integrated Space Transportation Plan) which, as he assumed the job of NASA administrator was muddled after the demise of SLI. As O'Keefe sought to advance the OSP something much more expansive was starting to take shape behind closed doors, the Crew Exploration Vehicle (CEV). While everyone thought that the OSP was the next big thing, O'Keefe, and a very small group of NASA employees and White House staffers knew that it was just the stalking horse for something much bigger.

But in order to see how NASA went from the OSP to the CEV, one has to look at all that NASA tried (and failed) to do under SLI.

X-33 – too many eggs in one basket

On 2 July 1996, Vice President Al Gore, with Dan Goldin at his side, stood before reporters at NASA's Jet Propulsion Laboratory to announce that Lockheed Martin had been selected to build the X-33 test vehicle. The X-33 was to be a half-scale model of an eventual Reusable Launch Vehicle (RLV) which would fly to and from orbit in one piece. Three teams competed for the X-33 vehicle: Lockheed Martin, McDonnell Douglas, and Rockwell International.

According to a NASA press release Lockheed Martin would, "Design, build and conduct the first test flight of the X-33 test vehicle by March 1999, and conduct at least fifteen flights by December 1999. NASA has budgeted $941 million for the project through 1999. Lockheed Martin will invest $220 million in its X-33 design. The X-33 will integrate and demonstrate all the technologies in a scale version that would be needed for industry to build a full-size RLV. The X-33 program is being conducted under a Cooperative Agreement, not a conventional customer/supplier contract. Under this agreement, NASA defined the broad objectives and industry proposed an approach to meet the objectives."

The X-33 was not just going to test one new idea—it would test dozens. A new engine, new thermal protection system, new internal structure, new ground processing and new guidance, navigation, and control. Moreover, unlike the way that the Department of Defense often constructs a new fighter with two companies competing, there would be only one X-33. If it failed NASA would have no alternative to turn to. This was a very risky endeavor, and these risks began to manifest themselves quickly as an optimistic project schedule fell apart.

The first flight of the X-33 was supposed to occur by March 1999. Less than a year after the project was announced it was already in trouble. In June 1997, developmental problems caused NASA to put off the first X-33 flight from March to July 1999. In October 1998 the first flight was delayed until December 1999 due to engine delivery problems. In January 1999 the first launch was delayed until July 2000 after a failure in a liquid hydrogen tank structure. In November 1999 a liquid hydrogen tank failed forcing the team to fall back to a heavier Aluminum Lithium tank design which was expected to push the launch date into 2001—perhaps 2002.

By the date it was supposed to have completed its test flights, December 1999, the first flight was still two years off. In October 2002, tests were pushed back another year, until 2003, five years behind schedule.

Undaunted, Lockheed Martin and NASA still pressed ahead with X-33. In late fall 1999, days after the devastating X-33 tank accident, Lockheed Martin was at the height of its promotional hype for the X-33. At the International Space Business Assembly in Washington DC, they were handing out X-33 shaped designer chocolates and mints. Not to miss an opportunity, they also staged a lavish event complete with a children's choir.

Staged in the Ronald Reagan Building's central atrium, a video showing the X-33 was visible on a giant screen followed by a large children's choir from the nearby Hoover Middle School who broke into repeated verses of a song titled 'Power of the Dream'. This whole presentation quickly became rather surreal with repeated flashes of Venture Star swooshing about on the screen while the children sang about dreams and their future. It was uncertain who was supposed to be impressed or motivated by all of this since the audience was comprised of aerospace company sales people, lobbyists, and NASA personnel. But it was exciting none the same, and seemed to herald the coming of a new era in space transportation.

The X-33 tank failure in the fall of 1999 led a number of people to call into question the entire strategy undertaken by NASA with regard to developing new launch technologies. In November 1999 House Space and Aeronautics Subcommittee Chair Rep. Dana Rohrabacher observed, "I am less hopeful about NASA's space transportation development efforts overall. Five years ago President Clinton transferred authority for developing reusable launch vehicle technology from the Department of Defense to NASA, despite bipartisan skepticism in Congress. Since then we have appropriated well over a billion dollars—more than the president has requested—to support these activities. But NASA's record of success has been poor by any measure."

In January 2000 Lockheed Martin agreed to put an additional $100 million into the X-33 project to either repair or replace the tank—so long as NASA did not end the X-33 program. But the handwriting was on the wall. On 1 March 2001, with a new administration in the White House to staunch the hemorrhage of money from the agency, NASA announced that the X-33 Program would not receive funds from the newly formed Space Launch Initiative. This decision more or less ended the X-33 program effective 31 March 2001 since Lockheed Martin chose not to pay for continued development of the X-33 with its own money. Indeed, Lockheed Martin had long since abandoned any notion that the follow-on, full-scale VentureStar could be built and operated on a wholly commercial basis and that government funds would be needed to build and then guarantee its operations.

All told, NASA claimed to have invested $912 million and insisted that it had managed to stay within its 1996 $941 million budget projection for the program. At the onset, Lockheed Martin had committed to invest $212 million. During the term of the project Lockheed Martin eventually contributed $357 million.

However these numbers were somewhat deceiving. According to an August 1999 GAO report 'Space Transportation: Status of the X-33 Reusable Launch Vehicle Program': "Under the terms of the cooperative agreement, NASA's contribution to the X-33 development program remains fixed, and Lockheed Martin and the industry partners are responsible for all cost growth. However, at least some cost growth may be recovered by Lockheed Martin and the industry partners by including the costs in their pricing for other government contracts. Further, costs for NASA personnel working on the X-33 program are paid from other NASA budgets. Thus, the government's share of the costs for the X-33 Program is greater than that represented in the cooperative agreement."

According to Congressional testimony by Allen Li, Associate Director of the GAO's National Security and International Affairs Division, a month later on 29 September 1999: "Estimated government costs for NASA civil service personnel working on the program also increased, from $95.2 million to $113.1 million. These costs for NASA personnel are not included in the agency's agreement contribution; they are paid out of another budget account."

Regardless of how the costs were parsed and reported in the final accounting, little was left to show for the effort. In the coming year debates would arise as to who got the remnants of the program. After a 6 month review interest by the Air Force in possible funding of the X-33 evaporated. All that was left to do was to argue over who got the scrap hardware. A mothballed X-33 launch site sat unused in California waiting for a rocket launch that would never happen.

X-34

On the same day that NASA walked away from the X-33 program, it did much the same to the X-34. The X-34 was to be a suborbital technology demonstrator launched from a L-1011 carrier aircraft and capable of speeds up to Mach 8 and would reach altitudes of 50 miles. The X-34 was intended to be a test bed for a variety of high reliability, low-cost technologies and operations for the next generation of NASA space vehicles. The X-34 was to be powered by a reusable 'Fastrac' engine, which was being designed and developed by NASA Marshall Spaceflight Center (MSFC).

Problems with the Fastrac continued to plague the program. The Fastrac was supposed to have broken new ground in the design, manufacturing, and maintenance of rocket engines by reducing the number of parts, among other things. NASA was totally insistent upon using this engine and more or less rejected out of hand a suggestion from Orbital Sciences and others that other engines already in existence such as Russian-built engines might be used to get the program back on track.

Powered flight testing of the X-34 was originally set for mid-1999 but slipped several times since the program began. New plans called for testing in late 2000 but this too began to slip. Talk of another 2-year delay was under consideration at the time the project was canceled. In addition, NASA considered a reduction in the number of powered flights from the planned 27 to perhaps as few as 2 to 6 flights. Moreover, the maximum speed would also be reduced from the planned speed of Mach 8 down to Mach 2.5.

With little hope that this project was going to provide a significant advance in knowledge (other than how not-to manage a program), NASA shelved this effort as well.

SLI – let a thousand flowers bloom

With X-33 and X-34 hardware being consigned to government surplus, NASA needed to show that it had learned its lessons and that the next attempt at solving launch issues would not take risk so much as it would beat it into submission. Whereas X-33 and X-34 embodied an approach where a large number of things were all tried at once, SLI sought to let a thousand flowers bloom. Hopefully, at some point, enough good ideas would emerge such that a launch system or two could be developed. The indications that something was amiss with X-33 and X34—even before they were canceled—had already prompted NASA to refocus its launch technology projects.

The Space Launch Initiative, with a budget of $4.8 billion through fiscal year 2006. was intended to reduce ("buy down") the risk inherent in developing the 'next generation or 'second generation' (the terms were sometimes synonymous) of NASA's launch capability. The core intent of SLI involved NASA working with the aerospace industry to design what would eventually become a privately operated second generation Reusable Launch Vehicle (RLV). Such an RLV would reduce the risk of losing a crew and reduce the costs of launching crew and cargo when compared to the space shuttle program. The logic being that if NASA could reduce launch costs and, at the same time, improve reliable access to space, that this would allow NASA to focus its efforts more upon its core missions.

NASA had hoped to be able to create at least two 'competing space transportation system architectures' by the middle of the first decade of this century such that they'd be developed to the Preliminary Design Review (PDR) stage and that they'd be ready for a full-scale development decision sometime shortly thereafter.

In a September 2002 report 'Space Transportation: Challenges Facing NASA's Space Launch Initiative' GAO urged a cautionary note as SLI proceeded: "NASA has to complete a reassessment of its overall space transportation plans. In doing so, it must decide whether it should continue pursuing the development of second-generation vehicles as planned, pursue alternative ways to develop the second generation in order to more quickly replace the space shuttle, or postpone these efforts altogether indefinitely until there is a major breakthrough in technology that could vastly improve performance and reduce costs. This decision will be difficult, given the uncertainties about the availability of technologies needed to reduce costs and enhance performance for future space flight." This caveat not withstanding, NASA had already made its decision as to what to replace – and when – and was pushing ahead.

GAO described SLI as "Part of a broader program—known as NASA's Integrated Space Transportation Plan—to address space transportation needs. Under the plan, NASA could operate the space shuttle through 2020 and make software and hardware upgrades to the shuttle in order to extend its use to this point. It envisions the deployment of second-generation cargo vehicles to begin around 2011 and crew vehicles around 2014. As with the shuttle, NASA envisions that the second generation vehicle will reach orbit in two stages. NASA also anticipates building a third generation of vehicles in 2025 and even a fourth generation in 2040."

In other words NASA would continue to use its 'first generation' reusable launch system—the space shuttle—until perhaps 2020 . It would do this while it developed the 'second generation' reusable system and then use both together. NASA hoped to define the basic requirements for the second generation reusable launch vehicle by November 2002.

X-37 /X-40: Overcome by events

One of the demonstrators selected prior to SLI managed to survive while X-33 and X-34 did not. The X-37 reusable technology demonstrator was designed to test a variety of concepts during spaceflight. Initially it was designed to be carried into orbit either by a space shuttle or atop an expendable launch vehicle. Boeing was awarded a contract to build the X-37 in 1999. The total value of the cooperative agreement,

January 28, 2003, Arlington, Cemetery. NASA Administrator Sean O'Keefe lays a wreath at the Space Shuttle Challenger Memorial. "NASA's success stands on the foundation of our unwavering commitment to safety, with respect to both our mission operations and daily activities," said NASA Administrator Sean O'Keefe.

January 14, 2004. NASA Headquarters, Washington, DC. President George W. Bush joined NASA Administrator Sean O'Keefe to announce the space exploration objectives for the agency. Veteran and active astronauts were on hand including Moon-walkers John Young and Eugene Cernan.

Photo: NASA/Bill Ingalls

President Bush is shown a full-scale mock-up of the Mars Exploration Rover during his visit to NASA Headquarters. The "Spirit" rover had successfully landed on Mars only ten days before the President's visit. January 14, 2004.
Photo: NASA/Bill Ingalls

January 14, 2004. NASA Headquarters, Washington, DC. President George W. Bush with an historic photograph, taken during the Apollo era, of the Stars & Stripes on the Moon. *Photo: NASA/Bill Ingalls*

January 14, 2004. NASA Headquarters, Washington, DC. President George W. Bush outlines his goals for NASA while Administrator Sean O'Keefe listens attentively. *Photo Credit: NASA/Renee Bouchard*

January 14, 2004. NASA Headquarters, Washington, DC. A host of active-duty astronauts—including "Teacher in Space" candidate Barbara Morgan (right)—applaud President George W. Bush as he announces his space exploration objectives for the agency. *Photo: NASA/Bill Ingalls*

President Bush meets with veteran Apollo astronaut Eugene Cernan just prior to making his announcement at NASA Headquarters. Cernan walked on the Moon during the Apollo 17 mission in 1972. *Photo: NASA/Bill Ingalls*

January 14, 2004, NASA Headquarters, Washington, DC. NASA Administrator Sean O'Keefe briefs news media representatives about the new vision outlined by President George W. Bush. *Photo: NASA/Bill Ingalls*

Vice President Dick Cheney confers with Dr Charles Elachi during his visit to JPL on January 14, 2004.
Photo: NASA/ Steve Benskin

14 January 2004. Vice President Dick Cheney is given a briefing at JPL about the
Mars Exploration Rovers *Photo: NASA/T. Wynne*

NASA Administrator Sean O'Keefe and Dr. John Marburger III, Director of White House Office of Science and Technology Policy present budgetary outlines. *Photo: NASA/Bill Ingalls*

February 12, 2004 Washington, D.C. House Committee on Science Hearing. NASA Administrator Sean O'Keefe and Dr. John Marburger III, Director of White House Office of Science and Technology Policy. *Photo: NASA/Bill Ingalls*

March 11, 2003 Washington, D.C. Senate Appropriation Subcommittee hearing. NASA Administrator Sean O'Keefe continues his testimony (right). *Photo: NASA/Bill Ingalls*

This artist's rendering represents a concept of possible activities during future space exploration missions. It depicts a human tended lunar base. *JSC2004-E-18883*

This artist's rendering represents a concept of possible activities during future space exploration missions. It depicts a crew with a large rover drilling on the lunar surface. *JSC2004-E-18831*

In a departure from tradition NASA have declined to release many graphics depicting the new initiative. This artist's rendering represents a concept of possible activities during future space exploration missions. It shows a Mars Artificial Gravity Transfer Vehicle. *JSC2004-E-18862*

including government and Boeing contributions, was to be around $173 million. This would be done in a 50/50 sharing fashion—with government funds coming from NASA and the U.S. Air Force. The X-40A, a smaller, 80% sized version of the X-37 was used for a series of atmospheric drop tests.

In 2002 the Air Force withdrew from the program beyond FY 2001 when vehicle weight issues undermined its value to the USAF. NASA continued ahead with the program and in November 2002, awarded Boeing a $301 million contract to continue the development of the X-37. Two X-37s would be built: one would be an atmospheric test vehicle, the other would fly in space.

Eventually launch aboard a space shuttle was dropped (due to the growing weight of the X-37 among other things) and the purpose of the X-37 morphed somewhat so as to be a pathfinder for the new, recently announced Orbital Space Plane. Indeed, during the summer and fall of 2002, when the configuration of the OSP was still only a rumor, one Washington DC-based industry employee was going around town telling people that he was convinced that the X-37 would be the OSP. Given that the X-37 was far too small to house a human crew—and certainly not one of the size envisioned for the OSP—this was incorrect. However, the mold line, the shape of the OSP design, Boeing later released did bear some superficial resemblance to the X-37.

By ummer 2003, X-37 problems were mounting with the airdrop program including some thoughts given to using a helicopter (instead of a B-52) for the drop test of the Approach and Landing Test Vehicle (ALTV). In addition, the weight of the orbital vehicle was now too great for flight on a Delta II such that a more powerful (and much more expensive) Delta IV or Atlas V would have to be considered.

Nonetheless, NASA went ahead and issued a pre-solicitation notice on 9 July 2003 which stated: "NASA/MSFC intends to negotiate on a sole source basis with The Boeing Company Phantom Works Southern California for a long duration orbital vehicle (LDOV) technology demonstrator under the Advanced Technology Flight Demonstration Vehicle (X-37) Project."

By Spring 2004, NASA interest in capsule designs for OSP and then CEV was growing, and interest in pursuing the original X-37 test vehicle beyond atmospheric drop tests was on the decline. NASA even looked at what it would take to recycle used Apollo capsules. Then, in a move some thought was an attempt to tempt the USAF back into the program, NASA revived an earlier requirement that the X-37 be able to stay in orbit for 270 days. Originally a USAF requirement, this had been more or less forgotten after the USAF pulled out of the program in 2002. NASA's rationale—officially at least—in reinstating this requirement was to test a variety of new remote-sensing technologies. Many thought this to be a thin excuse since there are many other ways to test space remote-sensing instruments.

The prime reasons were two fold: first, NASA was definitely moving in the direction of a capsule-based design for CEV. The X-37 has wings. Second, given NASA's preferred schedule, the CEV design process was to commence in earnest in Fall 2004. The first space flight of X-37 would occur no sooner than 2006. Since the core purpose of the X-37 is to provide data needed in the design of a possible follow-on vehicle, that data would arrive well after the CEV design process had been underway. As such, the X-37 would seem to offer little use to NASA in the design of the CEV.

X-43: Fast but no longer needed

In the mid 1990's NASA was looking at a variety of means to create launch vehicles that would be fully reusable – i.e. capable of taking off, delivering a cargo to space, and then return to Earth for refurbishment and reuse. Scramjet technology offered an attractive means to propel such a vehicle by lessening the amount of fuel carried. By accelerating to exceptionally high speeds air would be drawn into the vehicle and the oxygen combined by a propulsion system with hydrogen carried on board. High speeds were needed to deliver enough air to sustain the vehicle's engine.

NASA selected MicroCraft Inc., in January 1997 to construct three test aircraft. The X-43A program was planned to fly three X-43A test vehicles between January 2000 and September 2001 at speeds of Mach 7 and then Mach 10. Propelled by a solid rocket booster to great speed at which the scramjet could operate, the small test vehicle would operate for a few minutes, send back data, and then crash in the Pacific, their task accomplished. The first X-43A flight attempt in June 2001 failed when the Orbital Sciences Pegasus booster malfunctioned.

On 27 March 2004 the second X-43A did what it was supposed to do. Launched by a Pegasus booster dropped from a B-52 carrier aircraft the X-43A test vehicle's scramjet, ignited as expected and operated until its hydrogen fuel supply was exhausted. The X-43A eventually reached its target speed of Mach 7 and an altitude of 95,000 feet. Data sent back from the vehicle showed that the thrust developed by the X-43A exceed its drag and that the vehicle was accelerating when its fuel supply ran out. This was considered a rather major accomplishment for NASA. Only three X-43A test vehicles were built. The last one is scheduled for tests in October 2004 – possibly up to Mach 10.

NASA had planned follow-on vehicles. A larger version of the X-43A – the X-43C – was canceled in early 2004 as the new Office of Exploration sorted out all of the programs that it had inherited. The X-43C did not fit into NASA's plans any more. The X-43B, a larger test vehicle than the X-43A or X-43C, and the X-43D will never leave the drawing board.

Unlike all of the other X vehicles NASA had built in the past decade this one actually made it through the gauntlet past accidents and funding challenges, and went on to do what it was supposed to do.

Alas, given that NASA seems to be moving away from exotic space transportation systems and focusing on tried and true approaches, the X-43 will likely result in a treasure trove of data that will sit on a shelf for a decade until NASA is ready to use it. At a time when the agency's focus was shifting quite clearly to spaceflight—human spaceflight at that—many began to wonder if there was a future for advanced aeronautics research at NASA.

NASA renamed the Office of Aerospace Technology to the 'Office of Aeronautics' as part of the new policy rollout. However, its new Associate Administrator, Victor Lebacqz, did not sound all that confident about things in a briefing he delivered at a NASA Langley Research Center town hall meeting on 26 March 2004. Langley relies on aeronautics work for the bulk of its funding. On page 19 of his briefing, Lebacqz had a chart labeled 'Six Programmatic Priorities' for NASA's Aeronautics Enterprise. The last bullet on this chart read 'Determine if there is a requirement to continue hypersonics research'.

SLI starts to narrow its options.

At a 30 April 2002 press event, NASA proclaimed that it was, "Another step closer to defining the next-generation reusable space transportation system and successor to the space shuttle." The 'step' referred to a series of proposed concepts that had been put forth by the three contractor teams that had been selected—Boeing, Lockheed Martin and Northrop Grumman/Orbital.

Dennis Smith, NASA's SLI Program Manager, commented, "We said we would not pick a [vehicle] design at the beginning. We started with many concepts - hundreds - thousands of ideas. We whittled them down to 15 between these 4 companies. We will reduce this to 3 by the end of the year. We will then go into competition with 2 leading concepts chosen next year."

According to Smith, SLI was looking at a 60-70,000 pound cargo capability—this is more than the current Space Shuttle fleet is capable of carrying. Smith described the requirements of the SLI program as being focused to 'support the growing GEO satellite market'.

When asked if he could identify the specific commercial input (or reports) used by NASA to derive the SLI cargo capability (and if he would provide copies of these materials to the press) Smith replied that Futron Corp. had done a study. He then added that each of the participating companies had also provided input; but that this information was proprietary. When offered the chance to add to what Smith had said, all of the company representatives on the panel declined to comment. When pressed again to provide actual copies of the Futron report to the press, Smith said that he would. Weeks later NASA would allow the Futron report to be released. It was only four pages long and led many to wonder why such a short document would be the sole data point upon which to size an entire space transportation system.

Smith went on to say, "We're going to spend nearly $5 billion over the next 5 years to prove the capability to go develop the system." The time-frame Smith set was 2012 at which time a second generation launch system would be in place to provide crew and cargo transport to and from the International Space Station.

NASA's Office of Space Flight had recently begun a study to understand how to fly the Space Shuttle program until 2020. Regardless of whether these two dates remained firm, both the Second Generation Launch system and the current Space Shuttle system would be flying for a significant period of time. NASA had not been clear as to how these two systems would work together, as one was phased in and the other phased out. Moreover, this caused some concern among those who worried about how the current (Shuttle) system was going to be maintained.

Moreover, at that time, NASA was running a series of studies aimed at identifying potential business models for the possible privatization or commercialization of Space Shuttle operations. The process used would be competitive sourcing, a key component of President Bush's Management Agenda. Concern had been expressed on Capitol Hill that these studies did not have a firm model of what the Second Generation Launch System would look like as they went through their deliberations.

When asked about how the Shuttle and Second Generation Launch Systems would overlap—and if there was a requirement upon Second Generation architecture to do so (or if NASA had any firm transition plan)—Smith could not give a direct 'yes' or 'no' answer. Instead, he gave a long answer which, summed up, said that it is good that NASA is doing various studies.

The briefing closed with a question of where the funds for the development of a Second Generation Launch System would come from. "Do you have development funds in the 2007 budget?" Smith was asked. Smith answered the question partially saying that, "This is a decision that NASA's leadership is going to have to make," and, "we do not have the full cost estimates." Smith closed by suggesting that SLI efforts seek to raise 'all boats in the water'.

In other words there was no funding allocated for the actual development of whatever SLI eventually came up with, NASA had no official plan for how to operate the shuttle fleet (which might fly as long as 2020) and any future vehicles in an overlapping, integrated fashion, and it had already decided to replace the shuttle with a system at least as capable, based, by their own public admission, on a 4 page market survey by one company.

SLI's avowed aim was to make up for the mistakes made with X-33. X-33's VentureStar commercial follow-on was also a one-for-one shuttle replacement whose operation with the shuttle fleet was also given no serious thought. It would seem that NASA was not willing to give up anything nor learn any lessons. The issue of replacing the shuttle and how to phase the transition would be a problem NASA would soon have to face once again.

X-38 – Raising and lowering the pirate flag

One of the common headaches endured by the space station program was how to provide the ability for the crew to depart the station in an emergency and return to earth—possibly with injured crew aboard incapable of piloting the vehicle. During the Space Station Freedom Program an ACRV – Assured Crew Return Vehicle was spoken of, but funds for it were never included in the program. As Freedom was morphed into ISS the issue went on the back burner. Russian Soyuz vehicles would suffice for a while. This did not sit well with George Abbey, Director of NASA Johnson Space Center (JSC). Beginning in 1995, using center director's discretionary funds at first, and later funds from elsewhere, Abbey set up his own version of a 'skunk works' and found the perfect leader, John Muratore to run it.

Designed using previous USAF X-vehicle lifting-body research from the X-24A program, the X-38 program wisely did not seek to develop all new technology, but rather to utilize a wealth of research into previous spacecraft wherever possible. The X-38 would be able to carry up to 7 crew members in a reclined position and return autonomously to earth with a parafoil used to control its descent and landing.

The first X-38 drop test from a B-52 was on 12 March 2000 at Edwards' Air Force base and went smoothly. Successive tests would also run smoothly. X-38 progress not withstanding, there was already interest in alternate approaches to developing new modes of human access to space.

In a 3 December 2001 internal PowerPoint briefing to the X-38 team at JSC, X-38 Manager John Muratore noted, "We will all have to see what changes occur as the new administrator comes onboard. No one knows exactly what changes the administrator will want to make." He went on to say, "On at least four different occasions in the last year, I have been told that X-38/CRV was completely dead by a senior NASA manager. On each occasion, it has been reversed based on the merits of the project." Muratore then detailed a variety of programmatic changes which would allow costs to be cut, or actions delayed such that they would 'do the most to convince people to continue the effort to completion.' Muratore hoped that this would result in a slip for the first test flight of the X-38 until January 2005.

This was not to be.

On 13 June 2002 , NASA formally notified Congress that it was going to cancel the X-38 in favor of the Orbital Space Plane—something it now cited as being a multipurpose vehicle which would be capable of both transporting astronauts to the ISS and returning them—either in a routine or emergency situation. The X-38, by contrast, was only a contingency return vehicle.

Previously, at hearings on NASA's FY 2003 budget on 18 April 2002, Rep. Tom DeLay was furious at O'Keefe's move towards cancellation of the X-38 crew return vehicle telling O'Keefe, "You have a timid and anemic plan for human spaceflight."

In October 2002 Rep. Ralph Hall blasted O'Keefe's decision once again stating, "No quantitative analysis of the costs and benefits of X-38/CRV alternatives was conducted prior to the decision to terminate the program." His main complaint—as well as that voiced by others—was that the X-38 was ready to provide a crew return capability far sooner (2006) than the Crew Transfer Vehicle (CTV) capability for the OSP (equivalent to the X-38's capability) date of 2012 now set by NASA.

Nevertheless NASA went ahead with the shutdown of the X-38.

On Friday, 11 January 2003, one of four X-38 demonstrators was moved from the location at NASA JSC where it had been undergoing testing to another location where it was put into long-term storage. This was the V-201 vehicle which, had it been completed, would have been carried into space by a Space Shuttle Orbiter. It would have then made a fully autonomous re-entry and landing.

NASA decided to halt the X-38 program in 2002 once remaining test programs were conducted. This cancellation was part of a larger plan which emerged late in 2002 wherein NASA would proceed with an Orbital Space Plane capable of both crew transport and crew return missions from the International Space Station. The X-38 was developed to provide only a crew return capability from the ISS.

A photo posted on NASA Watch showed the vehicle on a flatbed truck as part of a small procession led across the JSC campus. The hummer was flying a skull and crossbones flag. Much to the chagrin of some NASA managers, the pirate flag has long been associated with the X-38 project. Indeed, as was the case with this move, it was often flown (with pride) to celebrate major project milestones—such as just after a successful drop test at Dryden Flight Research Center in 1998.

The pirate tradition stems from an earlier project led by X-38 project manager John Muratore—the installation of a new mission control center at NASA JSC. At that time Muratore's 'pirate team' adopted the pirate flag as their own. They even had a 'code of the pirate team' whose basic tenets focused on good teamwork while somewhat ignoring traditional NASA rules and management structures.

Ironically, one year almost to the day, NASA would re-discover the need for people like Muratore to spearhead some of its new exploration visions.

CCTV: Origins of the OSP

At the heart of the decision to cancel the X-38 program was the nascent OSP effort. As was the case with the X-38, the OSP openly reached back into a wealth of previous studies done by NASA and the military. Whereas the X-38 sought to eventually eliminate US reliance upon Soyuz spacecraft as interim ACRVs, the OSP sought to eliminate not only the Soyuz, but the crew transportation functions of the space shuttle fleet as well.

The requirements for this system can be found in 'Space Transportation Architecture Studies Phase III: Crew/Cargo Transfer Vehicle Preliminary Requirements', dated 22 June 1999, issued by the NASA HQ Office of the Chief Engineer. The Crew/Cargo Transfer Vehicle (CCTV) was an attempt to finally collect the requirements for a successor to the shuttle's various roles; requirements framed in the context of supporting a space station with a possible eye to eventually replacing the shuttle altogether.

Among the main objectives of this study was: "(a) To demonstrate an early complementary or alternate access system to the Space Shuttle for transfer of personnel and appropriate cargo to orbit and return. The CCTV system may provide additional personnel transfer capability during routine Shuttle operations or replace required Shuttle personnel transfer capabilities if the Space Shuttle becomes unavailable for any reason."

The baseline mission definition of the CCTV was: "The CCTV shall provide the following baseline mission capabilities when operating as a 'space taxi': An

International Space Station (ISS) crew exchange involving the transport, four times per year, of four crew to ISS (248 nmi and 51.6 degree inclination) and return. Small priority cargo carried either on the same flights, or on separate flights, may also be considered. In the event it is determined that the CCTV should also perform a crew rescue vehicle (CRV) role, the CCTV design shall be able to accommodate up to 7 passengers for emergency return from ISS in a manner consistent with CRV and ISS requirements."

In addition, the study included some 'expanded mission definitions': "Inclusion of other NASA transportation needs beyond those identified as baseline may impact the baseline CCTV design. CCTV design and operations shall be flexible enough to accommodate future increases in personnel traffic associated with increased ISS traffic and other NASA missions including potential high-Earth orbit (HEO) missions to support lunar, GEO, or Mars exploration missions. For example, the HEO mission requires a transport of 4 - 6 crew to a 100,000 km orbit and return and include rendezvous with an orbiting Mars transportation system. Also, there may exist a need to support personnel transfer and servicing at the L1 and L2 LaGrangian points."

Such capabilities would appear in partial form in the rationale for the OSP and in a full blown echo in the initial descriptions of what the CEV would be asked to do.

Morphing from one spaceship to another

At the time the CCTV concept was under discussion at MSFC it was often seen as political competition for George Abbey's X-38 at JSC. On the other hand, CCTV-inspired thinking spoke to some evolutionary uses of a vehicle like the X-38 which would initially be used as a crew return vehicle—possibly including upgrades to make it a crew transfer vehicle.

However, an unsuccessful attempt was made during the FY 2001 NASA budget cycle to start the phase out of the development of the X-38 and replace it with the CCTV. The CCTV concept had the backing of Steve Isakowitz who was at the Office of Management and Budget and would later move over to NASA as its Comptroller. Had the X-38 been canceled in 2001, the CCTV would have assumed the CRV's emergency return function as well as the ISS crew-carrying responsibilities of the current Space Shuttle system. In essence the crew return functionality of the X-38 would morph into part of what the CCTV was to do.

This effort was blocked by both internal and external political forces. George Abbey had expended a lot to get the X-38 this far and he was not going to give up without a fight. It would be more than a year before NASA Headquarters would be able to kill the X-38. When Headquarters did manage to kill the X-38 George Abbey had been removed from his job in February 2001. The CCTV concept would morph into what would become known as the Orbital Space Plane. Later, the Orbital Space Plane would also morph into what would become the Crew Exploration Vehicle.

OSP Emerges

The OSP had its origins in the budget amendment NASA submitted (later approved) in November 2002 which took the previously submitted FY 2004 budget and morphed it into a program to fix a number of ills at the agency. Under this shift in policy, NASA sought to implement a new Integrated Space Transportation Plan (ISTP). The existing ISTP was originally created as the policy underpinning of SLI. The SLI, in turn, was NASA's attempt to refocus its attempts to develop advanced launch technologies after the failure of the X-33 program.

The modified ISTP would now contain three major thrusts: the Space Shuttle, the Orbital Space Plane, and Next Generation Launch Technology (NGLT). SLI would remain in existence, but its focus would be tightened. The restructured SLI would contain the OSP and NGLT programs. Meanwhile, half of the funds originally targeted for SLI would be spent outside the confines of this program.

Key features under the new ISTP included developing an OSP to carry humans to and from orbit (and serving as a crew return vehicle), shifting the current decision date for Reusable Launch Vehicle (RLV) development from 2006 to no earlier than 2009, and continuing to fly the Space Shuttle until at least the middle of the next decade via enhancements to the Shuttle vehicles and the program's infrastructure.

With the after-effects of the Columbia accident echoing around the agency—plans to announce additional details of this new policy shift—as were to have been revealed in the FY 2005 budget the day after Columbia was supposed to land came to a grinding halt. After a few weeks to gather their thoughts, NASA decided to go ahead with plans to begin the process of defining the OSP. Some of the key aspects of the OSP now took on a new immediacy given that a second crew had died while flying a Space Shuttle mission.

Space Planes: not a new idea

The idea of a 'space plane' was not new. Looking at concepts developed for SLI one could see several examples included among proposed launch systems in 2001 and 2002. Even the name 'Orbital Space Plane' was not new. Astute watchers of the 1967 spy spoof 'Casino Royale' will recall that just as Woody Allen is about to dominate the world, a 60's coifed stewardess says, "Welcome to Dr. Noah's Orbital Space Plane."

The core idea is to launch a small winged spacecraft designed to carry humans and their luggage atop an expendable rocket. The Germans toyed with concepts toward the end of World War II which would have used a monstrous advanced version of its V-2 rocket technology. The USAF pursued the X-20 'Dynasoar' program in the 1960s which would have flown a winged crew transport vehicle atop a Titan rocket. The HL-20 concept developed by NASA Langley Research Center in the 1980s would have been launched on a Titan IV. Russian subscale prototypes and Europe's 'Hermes' concepts had similar attributes. Most vehicles never made it past the prototype phase before being canceled. Many never made it that far.

The only 'space planes' that did make it to space and then move on to approach becoming 'operational' were the American Space Shuttle and The Soviet 'Buran' shuttle. Both of which, while indeed being planes that flew in space, had a substantial cargo element to their operations—one that drove what the crew did—not the other way around. The Soviets abandoned their human/cargo Shuttle program when the expense to continue operations past one initial flight proved to not be worth the effort. It took America 30 years to make that same discovery.

Orbital Space Plane Revealed

On 18 February 2003 NASA released the Level 1 requirements for the OSP. At last the speculation was at an end. Although these specifications had been under development for months, the Columbia accident had added a new sense of urgency to replacing shuttles as fast as possible. Among the initial requirements:

"The system, which may include multiple vehicles, shall provide rescue capability for no fewer than four ISS crew as soon as practical but no later than 2010. The system shall provide transportation capability for no fewer than four crew to and from the ISS as soon as practical but no later than 2012. The vehicle(s) shall initially launch on an ELV. The system shall be operated through at least 2020. However, the system should be designed so that it could be operated for a longer time."

On 6 March 2003 NASA MSFC released an OSP Level 1 'Interpretation Document' which sought to clarify what the Level 1 requirements meant so as to allow the full fledged design requirements to be developed. But as was the case with the initial Level 1 requirements nowhere was there actually a specification that the Orbital Space Plane be a plane i.e. that it have wings.

NASA later awarded three study contracts totaling $135 million. Lockheed Martin, Boeing, and a Northrop Grumman/Orbital Sciences team each got $45 million. NASA asked these teams to perform certain technical engineering studies and to then further refine the Level 1 requirements.

NASA announced plans to hold a System Requirements Review in November 2003 and that it expected to make a decision on full-scale development of the OSP by the end of 2004.

Among the trade-off studies to be performed were to see if one vehicle or a family of vehicles would be needed; if the Orbital Space 'Plane' needed to be a winged vehicle or if a lifting-body or capsule design would be preferable, and whether the system or its components needed to be wholly reusable, partially reusable, or if they would best meet requirements if they were expendable.

Back to the Future?

Some considerable interest was mounting to return to concepts that had already been proven; i.e. capsules—possibly even non-reusable ones—so as to avoid the costly mistakes that went into making the space shuttle. Among the congressional supporters was Rep. Dave Weldon (R-FL) who said on 29 April 2003, "The capsule option has several other attractive elements. It would be cheaper and easier to implement an upgrades program. It could fly unmanned for test & evaluation, and also do re-supply missions. Flights could be more responsive and less costly with the elimination of the post-flight turn-around maintenance and reconfiguration that a reusable system requires. A flight certified capsule system that is proven to be robust, can be modified to go to the Moon or even as the return vehicle for Near Earth Asteroid or Mars missions."

While Weldon was speaking of the economics of flying humans to and from Low Earth orbit he let his personal biases slip i.e. that thought should be given to a system that would be capable of sending humans to places other than the ISS. Weldon's speculation was not idle, but rather well-informed since many at NASA were thinking along these lines and saw the OSP as a stalking horse for something much more capable than a space taxi.

Apollo revisited

Indeed, the day after Weldon spoke, an internal NASA study 'Report on Top-Level Assessment of Use of Apollo Systems for ISS CRV' was posted on NASA Watch. The report summarized a two day workshop held 13-14 March 2003. The membership of the team was dripping with experience from the early days of American human spaceflight.

On the team were: Aaron Cohen, former NASA JSC center director and former Manager of the Command and Service Module (CSM) in the Apollo Spacecraft Program Office, astronaut Vance Brand Apollo Soyuz, STS-5, STS41B, STS-35; Dale

Meyers former vice president and program manager for Apollo Command and Service Module at North American Aviation/Rockwell, former NASA Associate Administrator for Manned Space Flight and former NASA Deputy Administrator; astronaut John Young, Gemini 3, Gemini 10, Apollo 10, Apollo 16, STS-1, STS-9, and Kenneth Szalai former center director NASA Dryden Flight Research Center.

The team concluded unanimously that, "An Apollo-derived CRV [crew return vehicle] concept appears to have the potential of meeting most of the OSP SRV Level 1 requirements. An Apollo-derived CTV [crew transport vehicle] would also appear to be able to meet most of the OSP Level 1 CTV requirements with the addition of a Service Module. The Team also surmised that there would be an option to consider the Apollo CSM concept for a common CRV/CTV system."

The team did not suggest that museum hardware be used, or that an exact copy of Apollo systems be recreated. Instead, they suggested that the original design be used to guide a modern version which would have components with the same functions, perhaps even the same general shape, but capable of much more than was the original 1960s era Apollo system.

While no overt mention was made of missions to the moon or anywhere else, a paradigm shift best described as 'back to the future' was clearly the 'take-home message' this team sought to send out.

Regardless of how sexy the OSP could be, the notion of actually implementing an OSP faced two lurking hurdles: one was the space shuttle. Many people counted themselves in the 'Shuttle Hugger' camp and did not want to see this system go away any time soon. Others, notably from Texas, were angry over the cancellation of the home-grown X-38 project and wondered why the role of an OSP could not be filled, in part, by a revived (or expanded) X-38 program.

Doubts continue to linger

A few weeks after the report appeared, on 8 May 2003, in hearings before the House Science Committee, doubts about NASA's OSP plans continued to grow. Rep. Nick Lampson (D-TX) said, "NASA's proposed Orbital Space Plane program won't deliver a Space Station crew return vehicle until four years after we need it, and it will cost billions of dollars more than the X-38/CRV program that was canceled by the Administration. I think that's both shortsighted and wasteful. We can do better."

Subcommittee chair Rohrabacher said, "NASA views the Orbital Space Plane as merely a supplement to the shuttle. But it is unclear how NASA will pay to develop the Orbital Space Plane while operating the Shuttle, let alone whether NASA can afford to operate both the Orbital Space Plane and Shuttle at the same time. Additionally, NASA has yet to provide a clear picture regarding the strategy, schedule, and costs for the Orbital Space Plane."

One member, Rep. Joe Barton (R-TX), went far out on his own limb—and no one followed. Barton startled everyone by calling upon NASA to stop flying astronauts in the Space Shuttle altogether and to focus instead on developing new human launch capabilities that would be safer. "An accident rate of one every 62 and a half missions if 14 Americans have lost their lives is not acceptable. And it's my opinion that we can't make the existing orbiter as safe as it needs to be."

While no one else was ready to halt the shuttle program altogether, they were openly suspicious of NASA's latest plans. Rohrabacher would say, "I welcomed the restructuring of the Space Launch Initiative as a positive step towards making good on the promise of cheap, reliable, and safe access to space. As we begin to peel back the layers, however, NASA's proposed plan appears to be just another initiative that is long on promises and short on likely results. That simply won't cut it any more with this subcommittee."

As far as Congress was concerned, the OSP was yet another NASA idea thrown into the political process before all of the issues had been thought out.

Moving ahead

More or less undaunted by Congressional skepticism, NASA moved ahead with the OSP program. It released a Pre-solicitation Notice on 28 July 2003 stating its intention to 'negotiate additional scope under existing contracts, and solicit and consider proposals under a limited competition, only with the Lockheed Martin Corporation, The Boeing Company and a team consisting of the Northrop Grumman Corporation and the Orbital Sciences Corporation, for the design, development, test, delivery, and flight certification of an Orbital Space Plane (OSP)'.

Further, NASA stated, "This limited contractor group will proceed with the preliminary design of the OSP under the existing contracts, and will compete for the design completion, development, testing and delivery contract. The estimated period of performance will be through 2012." The reason for this increase in the pace of the program was also addressed, "The unusual and compelling urgency of the request is driven by the immediate need to develop an alternate human space transportation system for the International Space Station (ISS). The nature of this requirement is an acceleration of a previously anticipated 2010 Initial Operational Capability of the OSP. The objective of the potential acceleration enabled through this action is to provide initial operational capability of the OSP system as soon as practical, and the associated benefits to ISS utility and crew, up to two years earlier (from 2010 to as early as 2008 or sooner)."

NASA was apparently heeding concerns voiced by Congress that the time required to implement an OSP would leave the US dependent on Russian Soyuz spacecraft for Crew Return Capability. This was already a looming issue before the OSP notion hit the streets. Under the Memorandum of Agreement that guided Russia's participation in the ISS program, the Soyuz and Progress spacecraft provision ended in 2006. No agreement was in place to govern the provision of any more of these spacecraft beyond 2006. Given the Russian's proclivity to try and squeeze hard cash and concessions out of any and all interactions with the U.S., and the limitations imposed by the Iran Non-proliferation Act with regard to the U.S. purchase of goods and services from Russia getting these additional spacecraft was not going to be easy.

In September 2003 the OSP level II requirements were completed. Unlike earlier official statements by NASA, a hint was dropped in the Executive Summary of the Level II requirements as to what was in the back of people's minds. "The OSP will also provide a bridge to the future by serving as a foundation for future exploration missions." A press release accompanied the release of the requirements. The document provided an outline for the System Definition Review, which was scheduled for November 2003, as was a Request For Proposals which would be made to the three surviving contractor teams. "A decision to develop a full-scale vehicle system is expected in 2004."

On 29 September 2003, Lockheed Martin and Northrop Grumman announced that they had decided to enter into a teaming arrangement, with Lockheed Martin as the system prime contractor and Northrop Grumman serving as Lockheed Martin's principal teammate and subcontractor. This, of course left Orbital Sciences, who had been teamed with Northrop Grumman from the onset, out in the cold. This was rectified, to some extent, on 14 October 2003 with the subsequent announcement that Orbital Sciences had now been added to the team as a subcontractor.

What had looked like a three-way race to build the OSP had now collapsed into a competition between the two monoliths of the aerospace industry—Boeing and Lockheed Martin.

Wait a minute

A week later the House Science Committee had seen enough. Things were moving too fast and the ground upon which plans were being laid was not sufficiently stable. Rep. Sherwood Boehlert (R-NY) and Rep. Ralph Hall (D-TX) sent a letter to Sean O'Keefe.

Their complaint was rooted in the fact that they felt that, "NASA is proceeding with OSP development before we, the Congress, the White House, and NASA, have reached any agreement either on appropriate NASA goals for human space flight beyond the International Space Station or on the extent to which OSP is an appropriate approach to support those goals."

Specifically, Boehlert and Hall were concerned that the space policy review underway at the White House was still a mystery—to say nothing of what it was going to recommend. They didn't know, however, what CAIB findings were being discussed. And equally of concern were the Congressional issues, and those of others as well.

Their other concern was that, "Given NASA's current cost estimates for the program, the OSP five-year budget plan that accompanied the FY 2004 NASA budget request is clearly no longer credible." The cost estimates they referred to were those given by NASA's Dennis Smith in late September when he was making the rounds on Capitol Hill. Smith told staffers that the cost of getting to a CRV (crew return) capability for the OSP, by 2008, would cost between $11 and 12 billion. However, the cost to get the OSP to have a CTV (crew transport) capability atop an EELV was still not known. A strong case of sticker-shock, which was already developing among members, quickly grew. Smith's cost estimate caused a full blown case to develop.

Congressional concerns not withstanding, Boeing and Lockheed Martin pressed ahead. Several days before Boehlert and Hall sent their letter, Lockheed Martin was telling the Washington area aerospace community that it would be opening an 'OSP Demo Center' just south of Washington in Crystal City, Virginia. Among the things to do at this facility would be an OSP simulator and a video presentation narrated by Patrick Stewart (Star Trek's Captain Jean-Luc Picard). According to a statement provided by Lockheed Martin PR representatives, the OSP Demo Center "contains a full-scale simulated OSP cockpit. The center provides a place for decision-makers and stakeholders to visualize and understand the OSP concepts."

Putting the stalking horse on hold

As November approached, everyone was anxious to see the Request for Proposals issued for the OSP. On 20 November 2003 Rep. Boehlert addressed the Space Transportation Association, saying, "We're not, by the way, calling for a complete halt to the program or even for reducing the fiscal 2004 request, but we don't want to start taking steps that seem irrevocable. It's wrong to expect Congress to sign on to soliciting or awarding a contract for OSP when no one can tell us how the OSP fits into the future of NASA, or remotely how much the project will cost. You'd think Congress had learned that lesson by now."

Word quickly began to circulate that this would be delayed a bit. Then news emerged just before Thanksgiving that the RFP would not be released. NASA soon confirmed the news. No new date was given. Was OSP dead, or was NASA regrouping? With all of the White House space policy deliberations still kept under wraps, no one knew that the OSP had been serving for several months as a stalking horse for a much more expansive system—one linked to a policy that would send America back to the Moon.

That vehicle was the Crew Exploration Vehicle (CEV) which would not be announced for two more months.

The impression many had was that O'Keefe had bent to the will of Congress and put off the review until 2004. O'Keefe had put a halt to OSP development because he couldn't justify soliciting work on a system he was not going to develop. Since everyone seemed to think that NASA was regrouping, no one from NASA sought to explain what was actually going on. No one could, since only a few at NASA actually knew the full story.

Is there a moonship in our future?

The CEV was not exactly a surprise to many people when it was finally announced in January 2004. People were already flashing slides of things that could be assembled to form moonships. As the OSP concept had filtered out, many saw the possibilities inherent in an Apollo-style modular approach. Indeed, some of NASA's previous exploration planning had hinted at this sort of thinking within the tentative architectures that were occasionally discussed off the record. Hints dropped by NASA personnel that the OSP would soon morph into a 'moonship' or a 'Space Exploration Vehicle.' were becoming common just before, and immediately after, Christmas.

Indeed, at a gathering of space professionals in Washington on 18 December 2004, both Boeing and Lockheed Martin presented PowerPoint slides showing nearly identical plans for future space missions which bore a rough similarity to what would later be announced. Then again, revised Apollo architectures were not exactly news. Indeed, the presentations were so similar that either company's representative could have used the other's charts with no confusion whatsoever.

The artwork generated for the OSP concepts clearly hinted at thinking along more expansive lines. Boeing's OSP design (shown atop a Delta IV) clearly echoed Apollo—not surprising since Boeing had acquired Rockwell International which had acquired North American Aviation, the main contractor for the original Apollo Command Module.

Lockheed Martin first came out with a lifting-body concept (shown atop an Atlas V) and later revealed a second capsule design which superficially seemed to have both Apollo and Soyuz influences; with a little lifting-body flavor added. After the president's announcement, Boeing released a series of images which had clearly been generated some time before, which showed its Apollo-inspired CEV and a host of related modules and systems including a lunar lander and an inflatable crew module à la JSC's canceled Transhab design.

The CEV would be formally unveiled in President Bush's speech at NASA headquarters on 14 January 2004 confirming its rumored role and design inspiration. The president announced it, saying, "Our second goal is to develop and test a new spacecraft, the Crew Exploration Vehicle, by 2008, and to conduct the first manned mission no later than 2014. The Crew Exploration Vehicle will be capable of ferrying astronauts and scientists to the Space Station after the shuttle is retired. But the main purpose of this spacecraft will be to carry astronauts beyond our orbit to other worlds. This will be the first spacecraft of its kind since the Apollo Command Module."

OSP R.I.P.

Of the now fading OSP, O'Keefe would refer to efforts devoted to the OSP has having helped to flesh out a number of issues that would be directly relevant to the CEV. However, the OSP was dead. It would take another month, but NASA would eventually issue a modification to the original pre-solicitation notice that started off all of the OSP work by saying in officialese, "This is a modification to the synopsis entitled Design, Development, and Delivery of an Orbital Space Plane which was posted on July 28, 2003. You are notified that the following changes are made: This procurement is hereby canceled in its entirety. No final RFP will be issued, nor will proposals submitted in response to the draft RFP be accepted."

Marshall Space Flight Center had already put a lot into the OSP. Having been slammed hard over the past few years; first with the cancellation of the X-33 and X-34, and then by delays in Shuttle flights due to the Columbia accident. Some 800 people (500 contractor personnel and 300 NASA civil servants) had been working on the OSP at the time of its cancellation. People had been moved around, new offices established, and expectations raised.

While much of what went into the Level I and Level II requirements definition for OSP would indeed have applicability to the CEV, it would be perhaps a year or two before the CEV project would arrive at the point where OSP had been. As such, people were going to be reassigned or laid off, expertise groupings disbanded, and momentum lost.

Craig Steidle, head of NASA's new Office of Exploration Systems (Code T) would make repeated assurances that things would be O.K. for MSFC; but he never went so far as to say categorically that the CEV program would be managed in the same way, or at the same locations, as was the OSP program.

Looking back – and ahead

As such, if you were watching carefully through the years you could watch the X-24 and HL-20 morph into the X-38, then into the CCTV, then into the OSP, and then into the CEV. You would also see influences of the X-15, the X-20, Apollo, plus some hints of Soyuz and Progress. You would even see the hand of Von Braun.

Rarely does anything wholly new appear at NASA—or elsewhere in the world of spacecraft design. Rather, everything that flies is a derivative of something else, tweaked to push the envelope in a different, sometimes better way.

As is the case with bringing something new into the world, one has to adapt what already exists to accommodate that which is new. Given how hard it often is to bring new spacecraft into existence, many people are reluctant to let go of what they know and understand. But they also want the new things as well. As such, they try and have it both ways.

NASA's new space vision would come at a significant price. The retirement of the venerable, but flawed, space shuttle and an overhaul of the International Space Station. In exchange they'd get a new moonship.

Easier said than done.

*   *   *   *   *

8

CLOSURE AND DECISION

Keeping expectations in check

As fall 2003 changed into winter, the process of settling on the final policy and associated directives was well underway. Speculation grew. Word of an announcement that America would be going back to the moon surfaced in early December. These rumors soon contained a specific announcement venue and date: the 100th anniversary of powered flight on December 17th at Kitty Hawk. Few could argue that it was a natural place to make a major policy announcement about space.

Indeed, in a fall 2003 hearing, Rep. Dana Rohrabacher (R-CA) had suggested that this anniversary—and perhaps Kitty Hawk—might be an appropriate place to launch such a new policy. Few outside the White House/NASA inner circle knew the policy was more or less formalized, but was still in the final nip and tuck stage. While some thought was given to possibly announcing it at Kitty Hawk, this would only happen if it was ready to be announced. It soon became clear that the policy would not be completed, and the interest in aiming for Kitty Hawk evaporated.

Moreover, the Administration simply did not see the need to rush its own deliberative processes to meet external venues—regardless of the historic resonances they might offer. Eventually, the speculation was so rampant that the White House saw the need to dampen the speculation. Some vocal enthusiasts would apparently only be satisfied if they saw Bush walk out on the dais at Kitty Hawk and make an announcement. White House Press Secretary Scott McClellan addressed the issue. Several days later White House Chief of Staff Andy Card made mention on the Sunday talk shows— raising expectations a bit. In other words, 'Yes there is a policy. No it is not ready yet. When it is we will let you know'. Anyone who watched this White House should have recognized the response.

On 4 December 2003 Leonard David and Brian Berger wrote an article for Space News and Space.com titled 'White House isn't ready for the moon' wherein they sought to deflate some of the speculation that the Bush White House was indeed interested in space and formulating a policy to match that interest. Other articles followed and some of the speculation began to fade. At NASA O'Keefe was not at all unhappy to see the interest fade—even if the premise that his boss wasn't interested in space was incorrect. Less public arm waving about what might be going on would allow the real policy work to continue with fewer disruptions.

Around the same time this article appeared, former Congressman Bob Walker was telling people that he had information current as of 'last week' and that President Bush would be announcing a new air traffic management system at Kitty Hawk, "Because Sean O'Keefe can't get his paper work done in time." He added that, "Space policy will be rolled out in the state of the union." Walker's information was partially correct. It wasn't 'paperwork' that was the cause of the delay. Indeed, the finished documents had been passed to the White House the day before (December 3 2003) Walker was passing these hints. There was a new DOT/FAA/NASA/DOD air traffic management agreement that was completed and was available for use in case the White House wanted to announce something.

Several days later, an article by Jonathan Powell appeared at National Review Online which discussed a possible new Bush space policy. Nothing in the article was particularly new; all of it having been previously published on NASA Watch or other news outlets. What was different was the fact the Drudge Report had featured it thus waving it in the face of every assignment editor in the U.S. (even if they all deny visiting Matt Drudge's website). Within a few days the flurry by this article began to fade. By now, to his frustration, O'Keefe was being kidded by some about how he was 'keeping this quiet'.

Nonetheless, O'Keefe, and those involved in the final stages of the policy's development, did manage to keep the information under wraps. What swirled outside of the agency was usually fumes, with little substance. In order to help keep detail from being reverse engineered from other documents, the FY 2005 budget 'passback' to OMB (NASA's final response) was not circulated within NASA. This document was considered executive communication anyway and was usually treated as super-confidential. An extra layer was added this time. To have circulated it too widely within the agency would have tipped O'Keefe's hand, since the FY 2005 budget clearly represented the initial implementation of some major changes—changes that a sharp eye could ferret out.

The next week it became clear that although the policy was now in the final editing phase that it would not be ready for announcement on the 17th at Kitty Hawk. No firm date for its release was yet known and it could be any time in the next few weeks leading up to Christmas or the weeks after the New Year break. There was some urgency to get the policy announced as soon as it was ready so as to reduce the chance that it would leak.

On 11 December 2004 Vice President Cheney spoke at the dedication of the Steven F. Udvar-Hazy Center; the vast new annex to the National Air & Space Museum located adjacent to Dulles International Airport in Chantilly, Virginia. People were listening for hints as Cheney spoke:

"It is not by chance that so much of this history played out in the United States of America. At our best, Americans are a confident and a resolute people. When we set our minds to great objectives, we see the work through. The Air and Space Museum, both here and on the mall in Washington shows what can be accomplished with confidence, perseverance and unity of purpose. As the descendants of pioneers and immigrants, Americans are explorers by nature. And our native ingenuity and sense of adventure have been put to good purposes. Our air and space programs have been critical to the widespread prosperity of a continental nation. They've helped us explore space, not just for ourselves, but for the good of all nations. And in times of dangers, as in the war we're facing today, our mastery of aerospace technology has been essential to the success of our military and to the security of the American people."

Rumors flared up again on 12 December 2003 when an article by Marc Careau appeared in the Houston Chronicle which spoke of a new space policy and possible features including shutting down the shuttle program and rethinking the OSP as a moon vehicle. Careau was listening to the right drums in the distance. The aerospace contractor community was now hearing that the shuttle's days were numbered and Japanese company representatives were hearing of the OSP being reborn as the Space Exploration Vehicle.

Nothing happens at Kitty Hawk

As December 17th arrived, the president did indeed travel to Kitty Hawk to commemorate the 100th anniversary of powered flight. Not even the new multi-agency air traffic management agreement was announced. Space was mentioned, but in a context far broader than space exploration:

"A great American journey that began at Kitty Hawk continues in ways unimaginable to the Wright brothers. One small piece of their Flyer traveled far beyond this field. It was carried by another flying machine, on Apollo 11, all the way to the Sea of Tranquillity on the Moon. These past hundred years have brought supersonic flights, frequent space travel, the exploration of Mars, and the Voyager One spacecraft, which right now is moving at 39,000 miles per hour toward the outer edge of our solar system. By our skill and daring, America has excelled in every area of aviation and space travel. And our national commitment remains firm. By our skill and daring, we will continue to lead the world in flight."

While Bush had nothing overtly new to say, two people already in space did let drift a few tempting tidbits. Speaking from orbit in a live interview with CNN, Expedition 8 astronaut Mike Foale was asked about possible new space policy initiatives and whether the space station might be passed by for a direct push to send humans to Mars. Foale replied that he and crewmate Sasha Kaleri had spent "a lot of time looking down at Earth" and that "mountain climbing analogies came to mind." Foale noted that when attempts were made to climb peaks such as Everest and K2 that "a base camp was established along the way." He continued, "I see the ISS as a base camp to get to the Moon and the Moon as a base camp on the way to Mars."

As such, December 17th came and went and no grand space pronouncement had been made, despite hints from orbit. Now people began to wonder why. On 18 December 2003, Aviation Week's Space Imperatives Conference was held in Washington at the Ronald Reagan Building. A hundred or so attended. Among those in attendance was Apollo 11 astronaut Buzz Aldrin. Aldrin had been trying to pull together an event on the Kitty Hawk anniversary for some months. His original plans for a large gala at Kitty Hawk were soon scaled back, eventually morphing into this smaller event in Washington. Nevertheless, Aldrin managed to whip up a fair-sized 'who's who' of space and aviation policy. A similar theme was replayed at this event: everyone called for consensus but they left the meeting having achieved none.

Many in attendance suggested that a lack of an overt space policy announcement the previous day at Kitty Hawk was evidence either of indecision on the part of the Bush Administration or cold feet. Indeed they all felt that since they were all 'in the know' about such things, and that they saw nothing in the pipeline, that the space policy effort must be in trouble; otherwise, according to the prevailing view, they would all know what was going on. Absence of evidence was evidence of absence.

Quite the contrary. As these events transpired the policy was being drafted, written and waiting for a previously scheduled final run through by both the president and vice president.

One shrill rant during this period came from Robert Zubrin, president of the Mars Society. Zubrin had been a participant in Aldrin's event and had expressed his total dissatisfaction with the Bush Administration and NASA.

In one of his mass email messages titled 'Post Kitty Hawk Momentum Shifts to Mars' sent out on the day after the Washington meeting, Zubrin claimed that, "Human Mars exploration advocates emerged as the winners this week from the decision by President Bush to defer any decision about America's next major goal in space into 2004."

Zubrin continued, "In part because of the mobilization by the membership of the Mars Society demanding a real goal, and the right goal, for the space agency, the bandwagon for answering the Gehman report with a fake lunar program proclamation at Kitty Hawk was derailed. As a result, the decision process to determine NASA's new goal will now be prolonged until AFTER the Mars exploration rovers Spirit and Opportunity land on the Red Planet January 3 and 24, respectively. If either of the two Mars rovers should now land successfully, the extraordinary manifestation of public excitement about Mars exploration that will inevitably occur will make nearly impossible any new space policy that excludes human Mars exploration as its goal."

Alas, Zubrin's pronouncements were simply fantasy—perhaps the result of too many hallway conversations the previous day. The White House was firmly set on its own course—one perturbed only slightly by those in the space community who professed to 'know' what was going on—or claimed to have influence over what was being formulated.

With the long holiday break approaching, the chance that the space media, those most adept at finding things out, would have something to go with was becoming remote. Aviation Week and Space Technology only had a photo issue left in 2003 and Space News would be on hiatus until early January. Given the general ambience from such a large assemblage of space experts, the media were justifiably doubtful that anything was afoot.

Decision day arrives

When he arrived at the White House to meet with the president on the afternoon of 19 December 2003, Sean O'Keefe knew where the policy stood. Nonetheless he would be surprised by what transpired during the next hour. That Friday's meeting was the space planner's final review with the president and, if he so chose, the moment of his decision to proceed. The plan set before him that day had been built piece by piece in the methodical march towards consensus. It had started with the young space staffers thinking about space policy after the disaster in the skies above Texas. That moment had set in motion a chance alignment that so many in the space community had dreamed of for decades. The meeting had grown so large, and so many had been involved at the end that it required a larger room than normal. Those attending included Bush, Cheney, McClellan, the president's political adviser Karl Rove, Card and his special assistant Joel Kaplan, the president's science adviser John Marburger, Hadley and O'Keefe.

Rove had not been a big supporter of the idea and maintained a cautious attitude, although he did not criticize it. His silence was interpreted as support, for any critical word from Bush's most trusted political advisor would have sunk the plan or delayed it until after the 2004 presidential election.

The plan called for granting NASA an immediate—though relatively modest—budget increase, as well as an additional boost spread over several years. As Bush looked at the numbers, the others wondered if he would agree to them, given that only two other agencies, the departments of Defense and Homeland Security, were marked for increases in fiscal year 2005. Would the president agree and put his political capital behind the plan?

As the discussions moved toward a final choice—the Moon and then perhaps onward to Mars and beyond—Bush turned to Cheney. "This is more than just the Moon, isn't it?" the president asked. It wasn't so much a question as a statement.

Bush said he saw the policy as being more than picking a destination in space and then going there. Rather, it was more about going out into the solar system to accomplish a broader set of objectives. It should put to rest once and for all the decades-old and what one attendee called the 'somewhat tired argument' that space exploration was best performed by robots, not people. The new policy should embrace a mix of human and robotic missions, all focused toward a common goal: explore the vast reaches of the solar system; make that the centerpiece of American civil space policy.

There was a minute or two of silence. Then the vice president spoke up, "Then this is really about going to these other destinations, isn't it?" he asked. All agreed. One other item emerged: the president expressed a preference for inviting other nations to participate in the effort.

O'Keefe would later recall of this day, "The president was willing to put his prestige, standing, capital and reputation on this task even if it means there are critics, and that we are singled out as the one domestic agency with an increase in these challenging budget times. That instinct, frankly, has blown me away. It is a pleasure to serve in such an administration for such a man. These are the moments when this is all worthwhile … t'was a historic day."

Now that there was agreement on all the major points, Bush ended the discussion. "Let's do it," he said. The more the president had thought about the policy, the more he wanted to 'make a big deal about it' said one source that had attended the meeting. That decision was a surprise, since all assumed Bush would make a quiet rollout next Monday morning, 22 December 2003. Instead, he asked Hadley to find the next suitable date at which he could make a major space announcement.

6 January 2004 was open and was first chosen as the day to release the policy. Contrary to speculation by the media and people professing to have inside knowledge of what was going on, as soon as the final go-ahead was given to the policy, the next open date was selected. On 30 December 2003 the date changed again to 14 January 2004. This date was not chosen because it was after a Mars rover landing date. Nor was there actually any delay in coming to a decision, this despite speculation that missing one anniversary or another was indicative of indecision.

Now, as the year 2003 , NASA's annus horribilis, came to a close, the president was preparing to announce the new vision for space to the nation. It had been ten months and three weeks since Columbia had been destroyed. The civil space agency had moved from tragedy to renewal in the span of a single year.

On January 6th White House advance men and security would begin visiting NASA Headquarters to prepare for an event on 14 January 2004. The countdown had begun.

Hooray for the Deputies

Little known outside the councils of NASA and the Deputies Committee was the key role played by Steve Hadley and Richard Armitage. After all of the effort in crafting the policy had ended, O'Keefe would call them crucial players to NASA's success. Hadley had pushed the process along when it appeared to stall, or when things were on the verge of spiraling out of control as other events pressed in on the Bush White House. Margaret Spellings also had played a major role, along with Hadley.

And then, too, was the matter of Richard Armitage. Colin Powell's deputy at State stunned O'Keefe—and some on the White House staff not enamored of this whole idea—by providing a crucial bit of cheerleading for O'Keefe. "He was the enforcer," O'Keefe would recall later. Not just in regard to the international implications of the space plan, but also on the bigger, more expansive strategic vision for exploration. No one could explain how or why he became a space enthusiast, i.e. a true spacer, but surprisingly he had. "He really got hooked on this (the vision) and became the strongest ally we could ever have hoped to get," O'Keefe mused. "When in the annals of space policy history do you recall ANY Deputy Secretary of State advancing such an active position," O'Keefe asked.

When the process needed to be brought toward closure, Hadley had carried the president's dreams with him. He had the feel of what Bush wanted to get out of this process from the beginning. Now, at its end, he remained Bush (and NASA's) faithful steward, watching as the process exacted from NASA a very large price—reduced initial new money and an end to the shuttles and station—so that the exploration agenda could move forward. "Steve is the guy," O'Keefe would say, "that did the tough sledding. He defines 'multiplex',"" he added. Later, as a small way to acknowledge their crucial work, O'Keefe made sure the NASA Distinguished Public Service medal was awarded to Hadley. Marburger and Spellings.

A shot in the arm from Mars

As 2004 began, all was riding on a successful landing on Mars. While the White House had professed its scheduling indifference to specific mission dates, having good news in hand—fresh good news—as a harbinger of good things to come in close proximity to a major space policy was not lost on anyone. O'Keefe went out to Pasadena to NASA's Jet Propulsion Laboratory (JPL) with a large NASA entourage. Some of the members were logical; Ed Weiler, the Associate Administrator for Space Science and John Grunsfeld, NASA's Chief Scientist.

Others on the trip were less obvious. In an agency where human and robotic missions were often viewed as unrelated, O'Keefe was determined that there was going to be some cross-pollenation. Traveling with O'Keefe was Bill Readdy, Associate Administrator for the Office of Space Flight, manager of all human spaceflight activities. not a traditional guest at such an event. Ahead of his Headquarters team O'Keefe had sent—rather, had strongly hinted it would be a good idea— representatives from Mission Operations Directorate at NASA JSC led by Wayne Hale. Hale and his team were there as observers; seeing how the other half lives, so to speak. All of this was in keeping with O'Keefe's 'OneNASA' efforts whereby barriers between field centers and different disciplines would be taken down. By brute-force if need be.

Hale would later write a soul-searching memo that looked back at the Columbia accident and then forward, armed with lessons learned. This memo would have a profound effect upon many NASA employees.

Landings, human or robotic, can be nail biters. Less than a year before, in May 2003, only months after the Columbia accident, O'Keefe had experienced a harrowing landing vicariously. The occasion was when the crew of Soyuz 6S carrying the Expedition 6 space station crew, landed in the desolate steppes of Kazakhstan. After a long, anxious wait, as O'Keefe would later recall, he and the crews' wives waited as contact with the Soyuz crew was lost, and their exact landing site and status were unknown. Some time later the three occupants of the Soyuz were located.

Now, a $400 million robotic rover was going to land inside a bunch of inflated bags and bounce across the surface of Mars millions of miles away. It would then roll to a stop, unfold itself, and phone home, with pictures. Ample opportunity for more nail biting.

O'Keefe walked into the control room at JPL about 30 minutes before Mars atmospheric entry. He walked around, shook hands, and did the standard pep talk and then left. So far nothing all that unusual. O'Keefe then returned about 15 minutes later, to stay. Unlike his predecessor, Dan Goldin, who stayed out of the way until he knew there was good news, or to escape out a side door when the news was bad, O'Keefe's intent was to stick it out until there were cheers, or tears.

The original plan had been for O'Keefe to wait with all the other VIPs in the glass conference room adjacent to the Mission Support Area until good news had been confirmed from Mars. Only then would they send O'Keefe in to meet the team.

O'Keefe wandered over behind Steve Squyres from Cornell University, a geologist and the Principal Investigator for the twin Mars rovers. Squyres was sitting at his station, resting his head in his hands as O'Keefe walked up. He looked up and saw O'Keefe. Stunned, he said that he "couldn't believe" O'Keefe was there and "You can't believe what this means to everybody," i.e. that O'Keefe was there before the outcome of the landing was known.

About 5 minutes before Spirit landed, the entire control room was fidgeting. A voice called out announcing 'all nominal' as the reentry vehicle gained speed. Ed Weiler looked at O'Keefe with a very forlorn expression and said with a pregnant pause, "It's all over."

O'Keefe's heart nearly stopped. He gasped for breath and said, "Ed, what do you see that nobody else in this room seems to be noticing?" Weiler replied, "Oh, nothing remarkable. It's just that the landing has already occurred . We've got a 10 minute time lag on the comm, so we're watching the data that's already history. It's already either a success - or a disaster." Comprehension of what Weiler said took a second or two to register. O'Keefe would later joke, "My heart started again - and then I almost hit him!"

The rest of the night's events were seen by millions with O'Keefe hugging just about everyone in the room. Squyres would later tell O'Keefe that he could feel the tension in the room dissipate just a bit right after O'Keefe came in.

O'Keefe also departed from another tradition. Despite repeated requests from his staff that he don a jacket and tie, he refused to comply, and insisted upon wearing an MER mission polo shirt. As such, he blended in perfectly with the casually clad, overtly geeky mission managers.

O'Keefe's wife Laura told him later, "It's a beautiful thing to see nerds partying!" O'Keefe would later remark, "I can't believe the number of reporters who've observed that it's amazing to see scientists so excited. Isn't that what this is all about?"

Things start to leak

With bits and pieces of the story starting to leak out O'Keefe and others tried to get the announcement date moved up. Given the tightness of the president's schedule, and the fact that the president wanted to make a bit of a splash, a fixed date was required so that firm planning could be undertaken. Luckily for the administration, the Christmas and New Year's holiday break allowed the story to move to the back burner. Some publications were on hiatus. Civil servants were burning 'use or lose' annual leave, and reporters had no one to talk to, except each other.

A story that would likely have broken just before Christmas—didn't. The Administration had given some thought to announcing the policy in a less dramatic fashion at the White House the Friday before Christmas, but this plan was overtaken by other events and Bush's interest in doing something a bit more elaborate. Focus then shifted to the possibility of January 3rd or 4th. All of those dates slipped.

With the successful landings on Mars, everyone involved at least had the knowledge that there would be good news to follow; instead of bad news to explain.

When the January 6th announcement date started to take hold, O'Keefe had sent an email to Astronaut Mike Foale aboard the International Space Station informing him of the impending event and soliciting his participation in it. Foale was keen to participate. Had it happened on January 6th the event would have fallen on Foale's 47th birthday.

Eventually 6 January 2004 also fell off the president's schedule.

This date would soon change again. And now even President Bush was dropping hints. During a tele-conference with JPL employees on January 6th, JPL Director Charles Elachi invited Bush to come out and pay a visit to JPL. Bush smiled and said playfully, "Maybe we can kick around some ideas on quantum physics." The crowd roared. After the public part of the tele-conference ended, O'Keefe stayed on the line and ran the idea past Bush of having live crew interaction as part of the announcement event. Bush replied that he thought it was a great idea.

Still, every best effort was being made to keep the announcement under wraps until such time as the White House wanted to make it known.

During a press briefing on 6 January 2004, White House Press Secretary Scott McClellan sought to dampen speculation a bit by saying, "There's no update to what the president has previously said and what I have previously said, that, as you are aware, in the aftermath of the tragic Columbia accident, the president asked—or directed—his administration to review our space policy, and that is where it stands. The review has been moving forward, and I have no additional update at this time."

While the White House was doing its best to dampen speculation about any big announcement, later that day, White House Advance Teams visited NASA Headquarters to begin the process of securing it for a presidential visit. The date was now set, much more firmly than in the past, for January 14th, the day after Bush would return from a trip to Mexico.

On 7 January 2004 O'Keefe did a web chat on the White House website which caught the attention of a number of people. While the White House was trying to keep the lid on the story, other events, including O'Keefe's web chat and Bush's playful hints that something was ahead, had the opposite effect.

Later that day the story began to make its final attempt to break loose. Preparing an article that would appear in the following week's (January 12th) issue of Aviation Week and Space Technology, reporter Frank Morring had nailed many of the key details. NASA now knew that the story was building up steam and alerted the White House to the effect that something could appear very soon.

The next day other news services were starting to get hints as well. As the day progressed reporters could smell blood. After weeks of doing a dance to avoid answering questions with any specificity, White House sources now started to confirm long-standing suspicions that a space policy announcement was imminent. Late in the afternoon, based on persistent reporter inquiries, the White House began to acknowledge in somewhat vague terms that the president would be making an announcement on space policy the following week, no date was suggested, nor was any detail given as to what this announcement would entail.

Not wanting to wait any longer and be scooped, UPI went ahead with the pre-written story it had ready to go for several weeks. Titled 'UPI Exclusive: Bush OKs new moon missions' the story hit the UPI wire at 7:30 p.m. EST on January 8th. It then appeared above the fold on the front page of the Washington Times the following day. From that point on a feeding frenzy ensued.

Stories appeared elsewhere which either repeated details contained in the UPI story or containing comments from a number of people who should know what was going on but didn't. After a few days there was no additional information coming out about the actual substance of the policy. Three lengthy background pieces on UPI filled in many of the details, yet the final, official form that this space policy would take, and how the president would finally choose to present it, was still days away.

The UPI stories and other pieces generated nearly instant responses from Congress. It was clear by the speed of these responses, that people had been thinking of what they'd like to see in a space policy, and what they were going to say when the time arrived to do so.

Anticipation mounts

On 9 January 2004, House Science Committee Chair Boehlert said, "I'm eager to hear the president's vision for a revived human space flight program and I look forward to getting the details next week. I applaud the president for focusing on this issue at a critical time in the history of the American space program. I appreciate the meeting we had with Vice President Cheney in the fall, as well as my ongoing conversations with NASA Administrator O'Keefe and other Administration officials regarding the future of space policy."

Ranking minority member Bart Gordon said, "The president needs to provide leadership to place an ambitious human space-exploration agenda on the table, and Congress needs to be prepared to do its part. But I do not want to minimize the difficulty of doing what I am proposing.

"An ambitious presidential space agenda must represent a durable commitment, not simply one more re-election sound bite, or both Congress and the American public will dismiss it out of hand."

Similar sentiments came from the Senate. For a brief moment, everyone was paying attention to space. Congress had asked for a space vision. Now Bush was about to give them one.

By January 11th the text of the Presidential Directive containing the new space policy was done. On 12 January 2004 UPI broke the news that the event would taken place on the afternoon of 14 January 2004 at NASA Headquarters. Now the announcement had a bulls-eye painted around it, and everyone wanted to be there.

Announcement day

Control of attendees at the actual event was limited by several things. First the seating capacity of the James Webb Auditorium at NASA HQ, which would be diminished by White House staff, security, cameras, etc. In addition, all invitations were issued by the White House, regardless of whether NASA employees or other guests were concerned.

NASA had submitted a list of people it would have liked to have there, but the White House made the final call on all names. This helped NASA out a bit since it placed the onus on the White House to make the cut.

Nevertheless, NASA still received a number of phone calls and O'Keefe found himself constantly barraged with requests for a seat. Clearly, the number of people who felt they should be physically present vastly exceeded the room's capacity. The last time something of this importance occurred was in 1989 when Bush's father had stood on the steps of the National Air and Space Museum and proclaimed his own space policy. Before UPI had broken the date and time it had been somewhat of a hush-hush process to invite people.

Now that the date and the broad context of the announcement were known, the process of inviting people was simplified. But everyone who wanted to attend did not get in. On the day of the speech, NASA headquarters was virtually on a lock-down. Workers in their offices were not allowed to leave their floors. No one was allowed into the building. Around 1:00 p.m. the process of allowing invited guests into the building began. For the next two hours, the auditorium filled to capacity. Media were allowed in but had to stand in the back or sit on the floor. The upper seats were for NASA personnel and other government employees and guests. The lower seats, the first 5 rows or so, were reserved for VIPs.

POTUS is in the building

When the president arrived around 3:00 p.m. he was ushered into the building and up to the second floor. The day before, O'Keefe had talked to Bush by phone. "Tomorrow is your big day," Bush said. "Are you all set? I notice you had some edits to the draft remarks." O'Keefe assured him that all was set and ready to go.

As Bush and his security detail were escorted to a holding area, he was met by O'Keefe, O'Keefe's wife Laura, Ed Weiler, Bill Readdy, and Orlando Figueroa. Notably absent was Deputy Administrator Fred Gregory who was out on the west coast at JPL meeting with Vice President Cheney. Also present at Headquarters were a number of astronauts including John Grunsfeld, Ken Rominger, Ken Bowersox, Peggy Whitson, Ed Lu, Elaine Ochoa. Bush just walked into the room and said, "How are you all doin'?" and started to shake hands.

At one point Orlando Figueroa gave Bush an update on the Spirit Mars rover. Donning 3D glasses, Bush looked at photos that had been sent down from the landing site. Their conversation began in Spanish and continued comfortably for some time until they switched over to English. Figueroa then presented Bush with a highly detailed scale model of Spirit. This model now sits on Bush's desk in his private office. At one point, Bush turned to O'Keefe and asked what reaction he was getting from the folks on the Hill—that this is too expensive? O'Keefe replied that the reaction, "Went to both ends of the spectrum, some say too much, others say not enough." Bush replied, "Well, the objective is to find those three people in the middle who agree with this [policy] - that is what I do every day."

This visit to the agency's E Street headquarters was a first for an American president since NASA moved into the building in 1991. Bush's father had visited NASA's previous offices, and President Lyndon Johnson had visited NASA chief James Webb at the agency's first Washington home. Across the nation's small space community, in offices and laboratories with access to cable TV, work virtually came to a standstill. The rumors were over, now the truth of what had been in the works would be finally revealed. What direction would Bush attempt to take the country in space? Was it going to be a bold plan, matched with funding, or would it be an endorsement of the present? How could Sean O'Keefe find his way beyond the shuttles and stations to a new exploration agenda?

Showtime

With a packed audience of space and political dignitaries gathered in the auditorium, Bush finally ascended the stage at 3:25 PM. The president's speech itself was written by Mike Gerson, assistant to the president for speechwriting and policy adviser, using input from NASA and other sources. Around December 19th or 20th O'Keefe had submitted a rough draft covering the ideas that NASA had wanted to focus on. Gerson was also the person who coined Bush's poignantly simple message that, "The Columbia is lost." Spoken by Bush on the day of the Shuttle accident. Gerson also penned Bush's speech a few days later at the memorial service for the crew of Columbia. According to an October 2002 profile in the Washington Post by Mike Allen, "The hallmark of Gerson's speeches is the invocation of the vocabulary and literature of faith, and that only increased after Sept. 11, 2001." Gerson's hand could clearly be heard in some of Bush's passages.

Bush was introduced by O'Keefe and, by way of video hook-up, by astronaut Mike Foale aboard the International Space Station. Foale's insertion into the proceedings had been O'Keefe's idea, and as the timing of the speech was being planned, so were the opportunities for interaction with the astronaut. In fact, Foale had been among the handful of people, and the only one off the Earth, that had been clued in to what was coming for NASA. O'Keefe had emailed Foale, telling him only what he wanted him to do without giving him copious details of what Bush was to announce. But that wasn't necessary. By that point Foale, a personal advocate of a robust program of human exploration, had pretty much figured it out.

Bush began by referring to Foale's remarks. "I appreciate Commander Mike Foale's introduction," Bush said. "I'm sorry I couldn't shake his hand. Perhaps Commissioner…" Bush quickly caught his gaffe and referred to O'Keefe's correct title, "Administrator, you'll bring him by the Oval office when he returns, so I can thank him in person." He referred to the other astronauts, present in the auditorium. "I appreciate the astronauts of yesterday who are with us as well, who inspired the astronauts of today to serve our country."

Later in his speech Bush would refer to former Apollo 17 mission commander Gene Cernan who was present in the audience. Bush also acknowledged Rep. Tom Delay (R-TX), and Sen. Nelson (D-FL), who were sitting right in front of him. "I am honored that you all have come. I appreciate your interest in the subject," Bush joked, and Nelson laughed.

Bush began by laying out the complete package of what was called 'a renewed spirit of discovery'. Space exploration was, "A subject that's important to this administration, it's a subject that's mighty important to the country, and to the world," the president said. Inspired by 'all that has come before', and guided by the clear objectives of the plan vetted by the deputies' process, he set a new course for the space program. "Today I announce a new plan to explore space and extend a human presence across our solar system. We will begin the effort quickly, using existing programs and personnel," the president proposed. "We will give NASA a new focus and a new vision for future exploration," he said. "We will build new ships to carry man forward into the universe, to gain a new foothold on the Moon, and to prepare for new journeys to worlds beyond our own."

Embedded in the new space vision was a ringing endorsement of the value and place of space in the lives of American citizens. "We have undertaken space travel because the desire to explore and understand is part of our character," the president told the country. He listed more down-to-earth benefits from space-derived technologies. "Tangible benefits that improve our lives in countless ways."

Hailing the achievements of the space shuttles, he stated that their safe return to flight was his top, initial priority. The shuttles would be used to complete orbital assembly of the ISS. "We will finish what we have started," Bush pledged. But when that assembly was done, by no later than 2010, the era of the shuttles would end. Considering that in the years since the last Apollo landing in December 1972 no human had ventured farther out into space than roughly the distance between Washington, DC and Boston, Bush said it "was time for America to take the next steps."

It was a plan to explore space and extend a human presence across the Solar System.

He announced that following the shuttles would be the CEV, which he called "the first spacecraft of its kind since the Apollo Command Module." But unlike Apollo, Bush would initiate a program that would develop as few new elements as possible, making maximum use of existing systems, such as the expendable launch vehicle fleet. From the manned base on the Moon, the cost of missions to more distant locations might be reduced, by using the Moon's resources such as in creating oxygen or rocket fuel. Mars missions would follow the lunar expeditions, but the entire enterprise would be conducted in an incremental step-by-step fashion.

"We'll make steady progress, one mission, one voyage, one landing at a time," he said. But over and over in the speech, Bush repeatedly said that the Moon and Mars were just the beginnings of where he saw the space vision taking Americans. There would be ahead, he suggested, more distant travels. Now he added the technology elements that the staff had hammered out. "As our knowledge improves, we'll develop new power generation, propulsion, life support, and other systems that can support more distant travels. We do not know where this journey will end, yet we know this: human beings are headed into the cosmos."

Bush then laid out the budget profile. "Achieving these goals requires a long-term commitment," he said. Listing the space agency's current five-year budget at $86 billion, he told the nation that most of the cost of his new space vision would be paid by freeing up $11 billion from that total. But he would also ask for a budget boost as well. "I will call upon Congress to increase NASA's budget by roughly a billion dollars, spread out over the next five years," Bush explained. The budget increase, matched by the reprogramming effort, would provide NASA with 'a solid beginning' to meet the advanced space goals he had set. And he hinted that more money might be forthcoming if needed during a Bush second term. "Future funding decisions will be guided by the progress we make in achieving our goals," he added.

Bush turned to O'Keefe, who was standing silently and somewhat behind the president, and made it clear one of the reasons for the new policy's development; his family insider at the helm of NASA. "I am comfortable in delegating these new goals to NASA, under the leadership of Sean O'Keefe," Bush said, turning slightly to look at O'Keefe. "He's doing an excellent job."

In closing, Bush referred to the disaster that had started him on the pathway to this new exploration vision, the STS-107 accident. "We begin this venture knowing that space travel brings great risks. The loss of the Space Shuttle Columbia was less than one year ago." Bush noted that since the beginning of the space program, the country had lost 23 astronauts, and one astronaut from an allied nation. "Men and women who believed in their mission and accepted the dangers. As one family member said, "The legacy of Columbia must carry on; for the benefit of our children and yours.

The Columbia's crew did not turn away from the challenge, and neither will we."

Now, at the speech's end, Bush told the nation of his instinctive rationale for space exploration, the gut feeling that Steve Hadley had heard in Bush's comments during the space deliberations, and felt emerged whenever the president spoke of space. "Mankind is drawn to the heavens for the same reason we were once drawn into unknown lands and across the open sea ...we choose to explore space because doing so improves our lives, and lifts our national spirit." And then he was finished. "So let us continue the journey."

It was 3:43 p.m. on the east coast. Employees across the agency had followed the broadcast over NASA's TV system. It had taken 18 minutes to speak Gerson's paced cadence. It had been 11 months and nearly two weeks since Columbia had been lost.

Finished, the president moved down into the crowd of space believers, congressional members, astronauts, and NASA employees, stopping occasionally to have his picture taken. And then he was gone, departing an auditorium named for NASA's last champion of manned lunar missions, another master political manipulator who had no space experience.

Twenty-five minutes after he began to speak, his motorcade swept back out into E Street, cleared of traffic. In 10 minutes, the president was back in the Oval Office. In his wake, he left behind a changed landscape for space exploration. His vision encompassed nearly every dream ever offered by the space community.

Two hours later, out on the West Coast, Vice President Cheney spoke at the Jet Propulsion Laboratory. Accompanied by Deputy NASA Administrator Fred Gregory and JPL Director Charles Elachi, Cheney also praised NASA and referred to Bush's new proposed space effort, which Cheney called the second great age of space exploration. "These aims are ambitious," Cheney said. "They're difficult, and they're very demanding." But he said that if accomplished, the space vision would be repaid many times over, in "scientific advancement, useful new technologies, the discovery of resources on Earth and beyond, and the discovery of more about ourselves."

Cheney and Bush had confidence that the space program that they had laid out would succeed because, in part, "we have chosen exactly the right people to do the job." Elachi pointed to Bush's reference to a sustained program of exploration. "It's not a quick race to go to the Moon or Mars," Elachi said. It was a long-term program. A plan, he thought, that would highlight robotics on the moon to a degree never before done. Robots and humans, working together.

The vision settles in

This new space vision would come at a price. The space shuttle, the heart of NASA for more than three decades, would be retired. The space station, Ronald Reagan's space legacy, would be vacated in a dozen years. Nearly every other NASA program would be reshaped to support the plan. If it did not fit, it might well be jettisoned.

The firestorm over these tough choices was still to come.

The complete vision was specific about content but said little about execution. The U.S. would exit the space station around 2016, and would use its research time over the next 12 years in addressing issues related to keeping humans alive on other worlds. But what would happen to the station after the U.S. ended its primary use wasn't detailed.

Much of the early funding increases would go toward returning the shuttles to flight, the space station, and a new series of robotic lunar and Mars missions, beginning in 2008 with a lunar reconnaissance satellite. But it was not clear in what sequence and for what science the lunar robots would be developed. During the Deputies' process, Science Advisor John Marburger had urged that the Moon be the focus of a detailed scientific program, but the Bush policy appeared, at least at first blush, to value the Moon mainly as a jumping off point to the rest of the Solar System.

The development of the Crew Exploration Vehicle would be NASA's primary manned spaceship of the future, but scant details were available at the start of Project Constellation as to what the CEV would look like or what elements would be created to support its various missions. While Sean O'Keefe claimed that his agency wasn't 'wedded to a particular design', to achieve the Bush agenda within the tight budget timetables almost certainly would require a space capsule system; not that dissimilar from what was used in the 1960s.

While the CEV budget would be contained in the spending plan from 2005 to 2009, mission modes had not been selected. This would mean the 21st century Apollo-style assault on the Moon might use Earth orbit assembly, reducing the need for a new big booster. Such was Wernher Von Braun's initial idea for the Apollo architecture. With robots working alongside astronauts, new power generation systems, and new tools and spacesuits would be required.

Whatever pathway NASA would choose to achieve Bush's space agenda, a national commission, the 'President's Commission on Implementation of U.S. Space Exploration Policy' was to be established to vet the proposed plan. Soon known as the 'President's Commission on Moon, Mars, and Beyond', this limited term effort would be headed by former Air Force Secretary Edward C. 'Pete' Aldridge. The commission would be tightly focused on reviewing, not supplementing, the space vision plan, and reporting back to Bush by early summer. The 'Aldridge Commission' as it was also known, would exist for no more than 180 days and was chartered to provide its final report to the president within 120 days of its first meeting.

O'Keefe briefs the press

Less than an hour after Bush's speech, O'Keefe mounted the auditorium stage to speak to the news media. "This afternoon, we got a mandate," O'Keefe told the press. "And we got support for a set of specific objectives that very clearly identifies exploration and discovery as the central objective of what this agency is all about," he said. O'Keefe suggested that during NASA's 45 year history, much of those objectives and ideas had been central to its existence. "But to have it emphasized specifically as a reason in and of itself for these purposes is the important dimension of what this policy is all about."

O'Keefe did not say anything about the role he or the White House staff had played in shaping and reshaping that policy or how they sifted it down to the essential elements of human and robotic exploration that Bush had urged.

O'Keefe hinted that the NASA FY 2005 budget, to be unveiled in two weeks, would show the commitment to the plan in dollars-and-cents terms. It would "provide the assets, the resources, and the capabilities necessary to carry this out." Asked what specifically it would contain, he would only give vague generalities. "The five-year projection that's part of the budget release in a couple of weeks, again, provides specific program details that accomplishes that over time."

It would be nothing short of a transformation of NASA, a transformation that would begin with the status quo programs; the space shuttle and the International Space Station, then move on to encompass a new generation of robots and new spacecraft to sail out beyond Earth orbit again, as the agency had 30 years before.

"All of the inter-relationships between these factors will be built into this program for the purpose, again, specifically of pursuing the exploration agenda with the science to inform that set of goals as we move forward." O'Keefe said. And he promised that the transformation to come would be a reorganization of NASA from top to bottom. "We will organize ourselves in order to carry out that most efficiently," he said.

But how would NASA be able to do that? There were no specifics given. He did say that NASA would exploit new technologies to the maximum extent possible, not just in new spacecraft designs, power and propulsion but in management as well. That would be "part of our efforts in order to assure that we have the organizational means to carry out these objectives in a way that thinks about these challenges in the direction of large-scale systems integration, much as we did at earlier stages in our history," O'Keefe said. But he gave little specifics about how any of Bush's vision elements would be addressed. "Much more to follow in the days and weeks ahead," he said at the end of his remarks. "There will be a lot more discussion of the specific programs."

Transformation

As O'Keefe stood before the media and his agency that day, he knew that the months to come would be difficult, and that each and every day he would be called upon to move the ball toward the goal posts. In the months ahead, O'Keefe would use the word 'transform' quite often. When NASA did eventually announce the changes it was making in response to the Aldridge Commission's report six months later, it would become quite clear that O'Keefe had been thinking about a 'transformation' for quite some time. A transformation was far more substantial than a 'reorganization' in O'Keefe's mind. 'Reorgs' were something NASA would regularly do more often than a snake would shed its skin, with the impact being mostly to phone books, org charts, and business card printers.

O'Keefe knew that day in January that much more than a reorg was needed, and that he was going to use the momentum generated by the president's policy, along with the recommendations made by Pete Aldridge's team, to help make it happen.

But O'Keefe couldn't afford to wait for Aldridge to report back before making any changes. He needed to start shifting paradigms and mindsets right away. Along the way, O'Keefe would encounter roadblocks tossed in his path every day. Sometimes it was not possible to tackle every impediment. If he managed to make a convert once a day he considered that he had made some progress. A product of Jesuit schooling, O'Keefe would often refer to the process as making 'one convert at a time'.

Hitting the ground running

The next day, January 15th, the administration moved quickly to start restructuring NASA to support the vision. The agency announced the creation of a new exploration code T, headed up by a former Navy Admiral, Craig E. Steidle, known for his execution of the Joint Strike Fighter project, not any space program.

"Let us continue the journey," Bush had said, a journey that had begun in the fires of the now distant Cold War, and came nearly to a stop when that war appeared to have been won. Bush and O'Keefe would not have anything remotely like that war to help band together this new group of political supporters. Still, the achievement up to this point was impressive on its face. The long assembly of the political coalition within the executive departments had been achieved without sustained objection.

While investigators had built the answer to the Columbia's demise piece-by-piece, the administration had been assembling what could reasonably be called an entirely new path for the civil space program to take. While O'Keefe would eventually play the pivotal role in gaining Bush's support for the plan, the search for a new space policy had in fact predated him, and moved in its early stages at the White House without him.

The work of the informal 'Splinter' and 'Rump' groups laid the foundation of what the Deputies committee would ultimately build its consensus upon—return to the moon to prepare for an expansive exploration agenda, in essence to go elsewhere. Now the plan was public, and it was the task of Bush's political supporters in the Congress to fashion the next coalition—to get the vision assembled into legislative language. Such legislation would allow NASA to proceed with elements of the plan, especially fixing the shuttles and completing the station, as soon as the new fiscal year of 2005 began. That date, 1 October 2004 was less than nine months away as the president spoke. The other coalition would be of the public, for ultimately their support would be the most important element of all to sustain the plan.

As Friday, 16 January 2004 began, the atmosphere at NASA was ebullient. Everything seemed to be in place to roll out the space vision. But, around noon, just before what was to be a three-day federal holiday weekend, the first sign of trouble appeared. The team that had crafted the smooth rollout had left one deaf political ear aimed at an issue that wasn't supposed to be released to the public for weeks. But, suddenly, instead of basking in the glory of the new space vision, O'Keefe and his team found themselves under unexpected withering criticism.

The rollout was underway and under attack, from within the agency. No one had expected this.

*   *   *   *   *

9

SELLING THE PLAN

Packaging the Policy

As is the case with any new product, policy, or project, the success of a new effort—such as the Bush Administration's space policy—is a direct function of how it is initially presented. In Hollywood this is called packaging. Moreover, the continued success of such an effort also has a direct correlation to how well the plan for releasing the policy (the 'rollout') is constructed prior to release. Lastly, everything also depends on the flexibility of the plan to both anticipate events and adapt to changing or unforeseen circumstances as time goes on.

In the case of the new space vision, the rollout was more or less well-staged and garnered a lot of attention. However, the efforts in the days and weeks that followed were flawed from the onset. In addition, unanticipated events served to hinder the ability of the administration, and of NASA, to get ahead of the curve thus leaving the agency in a chronic state of being reactive, instead of proactive.

Congress complained that NASA provided them with insufficient detail to allow them to understand what the agency had planned—much less come out and embrace it. Advocacy groups were split along ideological lines and were slow to converge, and conflicting messages from NASA led the aerospace industry to become engaged in competing and overlapping attempts to support the plan. As winter morphed into spring and then progressed toward summer, NASA eventually managed to get its arms around these issues.

Yet at a time when NASA should have been culturing converts it was catching up and doing damage control.

A Self-inflicted Wound

Just two days after an unprecedented presidential announcement, and vote of confidence in NASA a bombshell dropped. NASA had decided not to send a shuttle servicing mission to the Hubble Space Telescope. Without this mission (SM4), the Hubble would likely suffer a premature end to its spectacular productivity.

The rationale for this decision had to do with safety issues for the shuttle crew, not with budget concerns. Moreover the decision was unrelated to the new space policy.

Much to NASA's dismay, the story did not break in a controlled fashion. Rather, it was individuals within the agency who sent out the details. In terms of formal responses, NASA was hours behind the curve, and caught totally off guard.

Here is how it transpired.

At 10:04 a.m. EST on 16 January 2004, an email arrived at NASA Watch stating "As of this morning, the Hubble program is being shut down effective immediately." At 1:29 p.m. that day a memo was received and immediately posted on the NASA Watch web site. This was a quick internal note from an astronomer at the University of Virginia describing events that had transpired an hour or two before.

"O'Keefe called a meeting of the 'Hubble Team' on Friday, January 16. We got the word 24 hours in advance. The meeting started at 11:30 a.m. at Goddard Space Flight Center in the Hubble conference room above the main bay (Building 7, Room 200 B/C) and ended at 1:00 pm. There were approximately 100 people in attendance, mostly from Goddard. Notable representatives included: Sean O'Keefe, John Grunsfeld, Ed Weiler, Anne Kinney, Eric Smith, Jennifer Wiseman (HQ) Steve Beckwith, Mike Hauser, Rodger Doxsey, Antonella Nota, Ian Griffin (STScI) Preston Burch, Dave Scheve, Dave Leckrone, Frank Ceppolina, Mal Niedner (GSFC) + a mix of around 100 people from Goddard and the contractors (I presume). NASA Administrator, Sean O'Keefe, delivered the news that he had decided to drop any shuttle servicing missions to Hubble in the return to flight, including SM4, and he wanted to tell us all in person about the decision. He said the decision was his alone."

Anne Kinney, Director, Astronomy and Physics Division in Code S (the Office of Space Science) at NASA HQ was also a recipient of the same email sent to NASA Watch. At 4:19 p.m. she replied with some corrections to the initial email stating, "Please allow me to correct an incorrect statement in your letter to the SOC. Code S did indeed identify funding to cover the SM4 slippage. But Code S did not oppose the decision of the Administrator. The decision was based, as you described in your letter, on safety. Code S fully supports the decision."

Damage Control

At 5:25 p.m. an email was sent out to the NASA press corps from Robert 'Doc' Mirelson, Chief of NASA News & Imaging stating, "We are doing a phone round-table tonite at 5:45 EST about the decision not to do another Hubble servicing mission. John Grunsfeld, NASA Chief Scientist will answer questions." Reporters called up but either could not get into the telecon or sat on the line waiting. The telecon was delayed as people were rounded up. Mirelson sent out another email saying that the telecon had been delayed 20 minutes.

After some additional technical difficulties, Grunsfeld was able to spend a short period of time explaining what was going on. It was clear that no one had planned ahead for the ramifications of this stunning announcement. To top off the lack of preparedness, NASA did not issue a press release that day. Nor would they ever issue one formally announcing the decision.

The storm began quickly and would not abate for weeks. O'Keefe would later reply to a number of inquiries by email saying, "… this was a very tough decision and not reached without a lot of prayerful consideration. Given our intent to comply with all the Columbia Accident Investigation Board recommendations for on-orbit inspection, repair requirements and contingency rescue requirements, a final servicing mission to Hubble would require a one of a kind set of procedures, training, and planning for shuttle flight operations before the shuttle is set to retire at the end of the decade. That poses a new set of risks which are surmountable, but significantly higher than station missions, the primary purpose of retaining shuttle for the next few years. Given the options for Hubble maintenance in lieu of servicing, we can concentrate on getting the maximum science from Hubble over the next few years before operations cease, so we will not abandon Hubble. Further, we need to accelerate the JWST deployment to the earliest opportunity the technical/production schedule can support; and that's our current plan and budget. When all those factors were considered, it was a tough, difficult decision to cancel the servicing mission. Rather than risk aversion, it's more evidence of prudent risk management. To be sure, the astronaut corps would have produced seven willing crew members with little reservation. While I can't expect you to accept or support that decision, please consider these factors and understand they were the elements dominant in my mind in making this tough call. Best regards, O'Keefe."

Loose Lips

What had happened can be traced back to a White House employee. O'Keefe had made the decision not to fly SM4 the previous month and had informed President Bush. Bush deferred to O'Keefe's judgment. O'Keefe had planned to announce it just before the release of the FY 2005 budget in early February. The reason being that it would be clear from the budget numbers that the SM4 mission was not going to be flown. O'Keefe wanted to have some distance between the policy announcement and the formal announcement about the cancellation of the SM4 mission. He also wanted to have time to talk with and prepare the Hubble family for this decision, and had planned to do so on 28 January 2004.

That plan was undermined a few days before the formal space policy announcement by the president. During a briefing on the budget to the House Science Committee, Gil Klinger from the National Security Council, inadvertently mentioned the topic. The rumors started to fly on 15 January when Kathy Sawyer put one sentence— "There will be no further servicing missions to the Hubble Space Telescope"—into a front page article in the Washington Post analyzing the president's policy announcement the previous day.

Realizing that the cat was about to be out of the bag O'Keefe hurriedly planned a meeting with the Hubble team.

While O'Keefe would later say that he had made the decision about the SM4 mission in late 2003, and that his decision had no budget or space policy linkage, the media, NASA employees, and anyone else paying attention to NASA and space policy made a direct linkage nevertheless. Proximity on the calendar between these two events had soldered them into one. As such, the first prominent repercussion in everyone's mind of this new space policy was the sacrifice of the Hubble Space Telescope. No amount of explanation by NASA would separate the two events in people's minds.

NASA's new space vision was barely two days old and it already had a big black eye.

The Hubble Family Strikes Back

A summary of the January 16th meeting by Space Telescope Science Institute Director Steven Beckwith described the mood in the room as being 'decidedly somber'. Determined to look for options Beckwith wrote, "On Monday, I am going to assemble a team at STScI to look at a wide-suite of options to extend Hubble's life. I think we should think positively about ways to use or even service Hubble without the shuttle and see if we cannot find something that would be attractive to NASA and preserve our observatory."

The lobbying effort to reverse the decision began almost immediately. Several websites with online petitions appeared almost instantly. A dry daily report of the Hubble's activities circulated on various Internet newsgroups—of which more than 3,500 had been issued—suddenly added the phrase, "Continuing to collect World Class Science" in its title. These reports also began to include alerts of media attention being paid to the SM4 cancellation decision. This was rather remarkable given the totally technical tone these reports had adopted for more than a decade. The Hubble folks were using every avenue available to them.

Long familiar with how to mobilize in support of a program or against an unpopular decision, the space science community sprang into action. In early February an article appeared in the New York Times which took issue with NASA's official stance that safety was the driver behind the decision. The article cited several anonymous documents which had been making the rounds.

One of the documents, titled 'Applicability of CAIB Findings/ Recommendations to HST Servicing' opened by saying, "The final planned HST Servicing Mission, SM4, will be at least as safe as shuttle flights to the International Space Station (ISS). If shuttle return-to-flight occurs prior to the full implementation of an autonomous inspection and repair capability, the overall risks during a flight to Hubble would be comparable to those associated with an ISS flight in which the shuttle failed to reach ISS."

The other document, titled 'Risk Considerations for HST and ISS Flights' stated that, "When looked at from a thermal protection system (TPS) survivability perspective, HST missions have a good basis for claiming that they are as safe as or perhaps safer than ISS missions conducted in the same timeframe."

While the actual author(s) of these documents was unknown, the documents were treated by the Hubble community, those who opposed the decision, and the media as irrefutable proof that the decision was flawed. The two documents, circulated as Microsoft word files, had the name 'wsmith' as the author. The Space Telescope Science Institute is staffed by AURA, the Association of Universities for Research in Astronomy. AURA's president is William S. Smith.

Clearly, anyone could have put that name in the document. When asked by NASA Watch to comment on this William Smith replied, "As you point out, the 'properties' function in the word document show that I opened then saved these documents. I did not however author them nor do I know who did. AURA was in the process of putting together an informational website to compile all available information. I had considered posting these but elected not to, due to the difficulty in attribution of such an anonymous document. The important issue, however, is not who wrote them but what they actually say. I believe they do state a position that I hope is taken into consideration in any final decision."

While Smith may not have written the documents, or known who the author was, he did make an effort to see that they were in the hands of those who might intercede. One congressional staffer, requesting anonymity, described how Smith had personally handed these documents out to congressional staffers. Senior NASA sources were also able to confirm this. As such, these documents had the de facto endorsement of the Hubble community and would eventually be linked to from the AURA website.

Arm waving

There were others who chimed in on the issue. In an article posted on Wired magazine's website Robert Zubrin was quoted as saying, "What's happening is that a bunch of bureaucrats are wanting to feel decisive, to show they can make the tough calls to support the president's Moon and Mars program. They'll say: 'Much as it might rend our hearts, we're willing to give this up.' That's all a crock. If the first thing this new space policy does is murder Hubble, then it's born with the mark of Cain on it."

Zubrin would continue on this rant for the next several months, dragging his entire organization along for the ride. Many scratched their heads wondering what saving the Hubble had to do with sending humans to Mars. Zubrin eventually condensed the core of his argument around the claim that NASA would not be able to send humans to the Moon or Mars if it was unable to mount a comparatively less risky mission in low Earth orbit.

Hubble Huggers in Congress Jump In

This was an issue that drew politicians in like a magnet. Most notable was Sen. Barbara Mikulski (D-MD) who guarded all space projects like a mother wolverine. In a 21 January 2004 letter to O'Keefe Mikulski claimed that she was "shocked and surprised by your decision to terminate the next scheduled servicing mission to the Hubble Space Telescope."

A week later CAIB Chair Hal Gehman was asked by Mikulski to review NASA's decisions in light of what the CAIB had recommended. Sean O'Keefe embraced the effort and said that he looked forward to what Gehman would have to say.

On 3 March 2004, a House resolution supporting a servicing mission for Hubble was introduced by Rep Mark Udall (D-CO). On 25 March, a resolution supporting a servicing mission for Hubble was introduced by Sen. Mikulski. Both resolutions called for additional studies to be done, but both fell short of explicitly calling for NASA to be directed to perform such a shuttle mission. No one in Congress, it seemed, felt comfortable in directing NASA to fly the missions, even though that was the obvious preferred end result. Rather, they hoped that by pushing enough people to re-think the decision that O'Keefe would submit to the pressure and change his mind.

On 9 March the Space Telescope Science Institute unveiled what it called "the deepest portrait of the visible universe ever achieved by humankind. Called the Hubble Ultra Deep Field (HUDF), the million-second-long exposure reveals the first galaxies to emerge from the so-called 'dark ages', the time shortly after the big bang when the first stars reheated the cold, dark universe." Of course, such awe inspiring, stunning imagery is what the Hubble had become famous for.

Speaking at the unveiling of the image, Sen. Mikulski said, "This is a stunning example of why the world loves Hubble, why I will continue to stand for Hubble. I will get the best minds to study the future of Hubble, for its future should not be decided by one man in a NASA back room, but by a transparent process. I know a second opinion is due but I want you to know I will not stop there."

Clearly Mikulski was going to keep the heat on NASA as long as she felt it was necessary to do so.

Gehman Replies

The next day Hal Gehman replied to Mikulski. Noting that the CAIB no longer existed he emphasized that the views he expressed were now solely his own. Many people were disappointed at the somewhat timid response Gehman presented. Instead of the crisp 'yes' or 'no' and pointed replies for which Gehman had become famous during the CAIB investigation, Gehman passed on a chance to backup the core recommendations that the CAIB had made, as NASA had applied them in the case of the SM4 Hubble Servicing Mission.

In addition to not answering the question asked of him, he punted on the issue at hand and suggested that further study needed to be done noting, "I suggest only a deep and rich study of the entire gain/risk equation can answer the question of whether an extension of the life of the wonderful Hubble telescope is worth the risks involved, and that is beyond the scope of this letter."

Curiously, the man whose panel pounded out an unrelenting mantra of 'prove that it's safe' now suggested in this first great test of the veracity of what the CAIB had recommended in his reply to Mikulski that NASA could go ahead and only 'do the best you can'. Indeed his letter served to lessen the stern tone of the CAIB's recommendations and allow more flexibility to be introduced into the process whereby NASA balances risk and benefit.

Many thought it curious that NASA would be the more conservative party in reacting to the first true test of the painful truths revealed by the CAIB report, and that its former chair would be more prone to compromise on safety issues.

On 11 March Mikulski responded to Gehman's report and asked the National Academy of Sciences to examine the issue. The Academy soon formed the

Committee on Assessment of Options for Extending the Life of the Hubble Space Telescope, chaired by Louis Lanzerotti, a familiar face on many space advisory panels in the previous several decades. The committee was charged to conduct an independent assessment of options for extending the life of the Hubble Space Telescope. The committee would study the risks inherent in a shuttle mission and what would be required to make such a mission safe; what robotic options were available, and options required for the safe disposal at the end of its useful life. When the committee would report back was uncertain, however such detailed efforts can take 6 months or longer to complete.

Out of Chaos Comes Opportunity

One thing that emerged from this decision was a renewed interest in robotic methods whereby Hubble might be serviced. In a teleconference held in January 2004 to discuss options for the Hubble, Office of Space Science Associate Administrator Ed Weiler had sought to cast serious doubt upon the notion of reboosting the Hubble. Weiler did so by saying that 5 ICMs (Interim Control Modules) "the system used to reboost the International space station," would be needed to accomplish the task. There was one problem. As Office of Space Flight AA Bill Readdy, also on the telecon, would add: the ICM was never completed and was not how the ISS was reboosted, contrary to what Weiler had asserted.

Over the coming weeks Weiler's tone, and that of his subordinates, would shift from outright skepticism to outright support for the notion of using automated means to extend the life of Hubble. This change in tone was the result of discussions between O'Keefe, Readdy, Weiler and others.

On 2 February 2004 NASA Goddard Space Flight Center issued a Request for Information titled 'Hubble Space Telescope End of Mission Alternatives'. Contrary to earlier interest by Weiler in only using robotics to deorbit the Hubble this RFI now sought "methods and technologies at Technology Readiness Level 5 and higher to extend the useful scientific life of HST and to safely dispose of the observatory at the end of the HST mission."

Contrary to the lack of official linkage between the new space policy and the decision not to fly the SM4 mission, the notion of developing robotic means to extend Hubble's life had a direct link to much of the initial thinking of how to implement the new space policy. Although NASA's space architectures were all still unofficial, many involved moving things around Earth, the Moon and elsewhere. One way to do this was with 'space tugs', automated propulsion stages that could grab onto, move, and then release satellites and cargo.

Some at NASA, notably Bill Readdy who had some previous hands-on experience with teleoperation research, saw the chance to use the Hubble situation as a stalking horse for some of the new automated space operation technologies that NASA would need to develop. Some arm twisting and direction from O'Keefe soon resulted in Weiler's remarkable transformation on this topic.

As such, while NASA had suffered a black eye (thanks in part to Gil Klinger at the White House) from its inability to properly orchestrate the SM4 cancellation, NASA was now able to transform some of the mess into an opportunity to get an early performance of the new capabilities NASA would need to develop as it implemented President Bush's new policy. At least all of the progress wasn't negative.

As 2004 progressed, NASA had moved ahead with its own studies and had focused in on a series of upper stages and robotic arms that it felt would allow the Hubble to be serviced and prepared for continued operations, and eventual disposal. The precise form these studies will take won't be known until late summer 2004.

Second doubts

On 26 May 2004 Sen. Kay Bailey Hutchinson upped the ante on the Hubble issue by sending a letter directly to President Bush which said, "NASA should keep Hubble operational by sending a manned space flight to perform the simple repairs and ensure the satellite's ability to provide crucial knowledge to our space science experts." To this letter was attached a petition signed by 26 ex-astronauts, ranging from Mercury pilots to late 1990's space shuttle veterans, which ended with the request, "we implore you to direct NASA to restore the Hubble Servicing Mission to the manifest."

On 16 June 2004, one of the petition's signatories, Astronaut Gene Cernan, sent a letter to Hutchinson. In it he admitted that he had agreed to allow his name to be affixed to a letter and that he did so "without participating as an author and without seeing a copy of the letter beforehand - regrettably breaking my own long-standing rules concerning the endorsement of someone else's agenda." Cernan went on to say that he had since received a detailed set of briefings on the topic and that he was "retracting my support as expressed in the subject letter. I believe we all realize the significance of both the scientific results of and public interest in the Hubble. I also believe that if there is a way either manned or robotically, the Hubble will continue to be a serviceable asset without comprising the shuttle's primary mission of space station completion."

Another astronaut, Lt. General Thomas P. Stafford, Co-chairman Stafford/ Covey Return to Flight Task Group, who had not signed the letter, wrote a letter to

Hutchinson defending NASA's decision on the SM4 mission saying, "It is not necessary to send humans when a robotic mission can extend the useful life of the HST. By advocating an unmanned robotic mission to the HST, NASA has not only embraced the CAIB recommendations, but has taken a step to preclude undue risk. As an added benefit, the demonstration of such robotic capabilities will be important as NASA implements the vision for space exploration."

This issue would be a hotly debated topic by a number of the petition signatories on 22 June 2004 in Houston at a meeting of the Association of Space Explorers. ASE is a rather exclusive organization which requires that its members have at least one space flight under their belts. The ASE did not take a position on the petition, but had made the effort known to its members.

Limping Ahead

As a result of the Hubble mess NASA was forced to waste precious time and forcibly evaporate much of the good will in Congress, and elsewhere; good will that could have been used to move the president's new space policy ahead. Everyone was distracted by an avoidable mess.

When NASA did get down to the task of moving the new policy through Congress it did so with one hand tied behind its back: its Office of Legislative Affairs was in disarray. In November 2003 it became known that Associate Administrator for Legislative Affairs Charles Horner would be leaving NASA. He would be replaced by D. Lee Forsgren from the law firm of Adams & Reese - the same law firm NASA General Counsel Paul Pastorek once worked at.

Forsgren was new to the topic of space, but not to Congress and politics. He had worked on the Transportation Advisory Committee for the incoming Bush-Cheney transition team. Previous to that he had served for five years as the Majority Counsel for the House Committee on Transportation and Infrastructure's Water Resources Subcommittee. Forsgren also served as Assistant Majority Counsel on the House Coast Guard Maritime Transportation Subcommittee and as the Maritime Legislative Assistant in the office of Rep. Don Young (R-AK).

Horner, a Democrat and decorated Vietnam combat veteran, was a long time friend of Sean O'Keefe's and had joined NASA in May 2002. Before O'Keefe had picked him to come to NASA, Horner was the Principal Deputy Assistant Secretary of the Army for Financial Management. Before that Horner was a special assistant to the Secretary of the Navy, and a special assistant to several DoD comptrollers.

Despite his credentials, it did not take long for Horner to make his mark on Capitol

Hill. Frequently described with terms such as 'bull in a china shop', Horner soon generated a significant amount of ill will. He managed to help O'Keefe pull off some notable achievements including a reprioritization of the already approved FY 2003 budget in late 2002 to jump-start some of O'Keefe's initiatives. Horner was soon making more enemies than friends. In addition Horner inherited a staff which had been reduced in effectiveness by attrition and combat fatigue after a decade under Dan Goldin.

Given O'Keefe's deep background in congressional process, one derived from many years as a staffer himself, many were becoming increasingly mystified as to why O'Keefe had kept Horner on, and why NASA's Congressional relations were not on the upswing.

Known for his fierce loyalty to those who work for him, O'Keefe eventually had to make a very difficult choice with regard to Horner. Looking ahead to the even larger Congressional battles he'd have to win as the new space policy hit the street, it was time for a change.

Promoting him out of his job to be a special assistant, O'Keefe would later privately say of Horner, "He is the best ninth inning closer I have ever met. The problem is that he does not know how to start the game or play the middle innings." O'Keefe now admits that putting Horner into this role just did not work out. "This was a big mistake," he would say. Clearly fond of Horner O'Keefe added, "But he meant well."

When word of the impending new space policy began to solidify, many wondered why O'Keefe, who had been planning this for some time, had not replaced Horner much sooner so as to have a seasoned veteran in place, ready to go. O'Keefe notes that he wasn't able to do this since Forsgren could not leave his previous job until the end of 2003.

To be fair, NASA's Office of Legislative Affairs was already showing some cracks under O'Keefe before Horner arrived. Horner replaced the previous AA Jeff M. Bingham. In April 2002 Bingham, who had once worked for Sen. Jake Garn (R-UT), had allowed an embarrassing confrontation between O'Keefe and Rep. Tom DeLay to occur; one that could have easily been avoided.

At hearings on NASA's FY 2003 budget on 18 April 2002, DeLay took serious issue with O'Keefe's management of NASA. DeLay, who now had Johnson Space Center in his district after a contentious round of redistricting in Texas, was furious at O'Keefe's cancellation of the X-38 crew return vehicle—a home grown JSC project. While he had begun his comments calmly by telling O'Keefe, "It is nice to see you again." DeLay soon went for the jugular and told Sean O'Keefe, "You have a timid and anemic plan for human spaceflight."

O'Keefe's response was, on the contrary, confident and measured. In addition to telling DeLay, "It was nice to see you too," O'Keefe threw DeLay's words back at him several times saying that what he was doing was "hardly a timid and anemic plan." DeLay's face was beet red at the end of the exchange.

Hours after the confrontation, O'Keefe sent NASA comptroller Steve Isakowitz over to Delay's office as a peace offering. DeLay let Isakowitz sit and wait in his outer office for over an hour, and then had someone tell him to make an appointment another day.

Both sides were sending messages.

When questioned about the exchange with DeLay the next day at a reporter's roundtable at NASA HQ, O'Keefe said, "So long as there is no ambiguity we can deal with the issues honestly." With regard to DeLay, O'Keefe admitted that he needed to "redouble his efforts so that communication flows to friends and critics alike."

Bingham was replaced by O'Keefe three weeks later.

Forsgren showed up for work at NASA during the first week in January having been a consultant for a short period before that. What he walked into was a massive new space policy that was 99% complete and ready for implementation. The problem facing Forsgren was not simple. In addition to the steep learning curve he had to mount, his office was in disarray. By now most of the talented employees had moved out of the agency. Those who remained were long time veterans who had lost the spark that one needs to promote something new and exciting.

Congress Chimes in

The day after the announcement, reaction was predictably mixed. The San Francisco Chronicle quoted Rep . Zoe Lofgren (D-CA) as saying, "It clearly is massively insufficient to advance the program that he is touting … It makes me wonder whether this is just another political plan that really has nothing to do with reality."

Former Vice President Gore was openly dismissive. According to Reuters, on 15 January 2004, "Instead of spending enormous sums of money on an unimaginative and retread effort to make a tiny portion of the moon habitable for a handful of people, we should focus instead on a massive effort to ensure that the Earth is habitable for future generations," Gore said to a cheering Manhattan crowd." Howard Dean was no more charitable according to the McClatchy Newspapers, "Former

Vermont Gov. Howard Dean, campaigning for the Democratic presidential nomination in New Hampshire last week, said Bush's plan 'is not worth bankrupting the country' and suggested it is politically motivated. Some analysts said Bush wants to inject a bold, forward-looking domestic initiative into his re-election campaign this year."

Reuters quoted Sen. Joe Lieberman (D-CT) then still a candidate for the Democratic presidential nomination as saying "We should not be going hundreds of millions of miles away on a costly new mission when we have limited resources." And the man who would come to be the nominee of his party, John Kerry was quoted by the San Francisco Chronicle as saying, "Rather than sending Americans to Mars or the moon right now, these people would be better off trying to figure out how to get Americans back from Iraq."

Sen. Bill Nelson (D-FL) was supportive but still skeptical according to the Houston Chronicle, "But Sen. Bill Nelson, D-FL., who flew aboard the space shuttle in 1986, said the additional $1 billion Bush proposes isn't nearly enough to accomplish all the plans. 'If the president will lead, Congress will support it,' Nelson said. 'But he's got to put the juice to it.'"

Rep. Dave Weldon (R-FL) was far more supportive. The Orlando Sentinel wrote, "Some may say now is not the time to talk of going back to the Moon and on to Mars. I say now is the perfect time. Failure to move forward on these types of initiatives means failing our next generation. When we commit to exploration today, we expand the horizons and possibilities for our children tomorrow. That's why the president's announcement is good policy and good leadership."

Way Out Comments

Of course, there were those who went beyond rational discourse and tip toed on the edge of the bizarre. Andy Rooney, CBS' resident curmudgeon said, "But then, I think of all that money. There are so many things we need to do here at home. Space exploration hasn't produced much for us except some good pictures." Disgruntled former NASA historian Alex Roland told the Orlando Sentinel, "I think it's just a circus. All the arguments that are made for it are that humans have to be there to explore. I ask them why, what is it that people do there, that makes it worth the enormous cost and risk? And they get all vague and fuzzy at that point."

Perhaps one of oddest claims was made in an online article by Wired magazine about the policy and the Hubble servicing mission cancellation, "Space policy analyst John Pike of GlobalSecurity.org sees dark motivations behind the move. He argues that President Bush has made clear what will be abandoned in the short term, but has scheduled non-military missions like colonizing Mars far over the political horizon so that succeeding administrations can cancel them without controversy. 'I think it is sort of symptomatic of this administration's orderly dismantling of the American space program,' he said."

Clearly not wanting to be accused of being unsupportive, presidential candidate Dennis Kucinich (D-OH) issued a statement in early March saying, "Kucinich, co-sponsor of the Space Exploration Act of 2003, said the current budget for NASA "is far from adequate. Our shuttle fleet is based on 30-year old technology and this is only because of a lack of funding. Although the shuttle program requires $4 billion a year to operate, NASA has been forced to operate the shuttle with a budget of only $3 billion a year."

State of the Union ?

When the initial rumors were swirling in late 2003 and the first days of 2004 about the venue where a new space policy would be announced, the State of the Union was one of the common suggestions. That speculation more or less came to an end when word came that there would be a special event at NASA headquarters featuring the president. When the speech was given on 20 January 2004 - less than a week after the space policy rollout at NASA Headquarters, no mention was made of the new initiative in the speech.

A week later, in Hearings on the new space policy before the Senate Commerce Science and Transportation Committee, Sen. Nelson asked Sean O'Keefe why there had been no mention of the new space policy in the speech, noting that without strong support from the White House Nelson was "afraid it's going to fizzle." Adding, "When the president announced this last week, I was one of his biggest cheerleaders. Only one week later, lo and behold, it looks like they dropped it like a hot potato."

O'Keefe did his best to try and downplay the specific mention of the initiative in the speech and that the White House was fully behind the plan. But the skepticism lingered. According to Reuters Sen. John McCain (R-AZ) was also skeptical, "I think the American public is justifiably apprehensive about starting another major space initiative for fear that they will learn later that it will require far more sacrifice, or taxpayer dollars, than originally discussed or estimated."

On 28 March 2004 The New York Times published an interview with President Bush's chief communications strategist and confidant Karen Hughes about her new book 'Ten Minutes from Normal'. Mention was made of the absence of space in the speech, "Ms. Hughes also advocated dropping from the State of the Union address any mention of the president's ambitious proposal to send humans to Mars, which was attacked by members of his own party as extravagant folly. 'At some level the policy gesture didn't pass the communications straight-face test,' the Bush adviser said."

Hughes' recollections struck many in NASA and the administration as being odd. On one hand, the reason why the new space policy was not mentioned was the same reason that many other things had not been mentioned, they did not fit into the messages the president felt he needed to put forth that night, and he only had so much time to mention the things he thought needing mentioning. The White House also felt that the policy had received quite a fair amount of focused attention several weeks earlier when it was announced, with some flourish, and that it was felt that it was time to let the initiative work its way through the political process.

Where's the momentum?

Although no one at NASA will openly admit it, many thought that Congress, which had been so vocal in its call for a space policy, and a space vision, would flock much more quickly to the ranks of the supporters of this legislation. Veteran space policy analyst Howard McCurdy, from American University put it best when he told the Orlando Sentinel on 8 March 2004 that NASA had hoped a key legislator "would stand up and raise the flag and say 'Let's go! Follow me!' And that hasn't happened."

The next day, House Science Committee democrats released their 'Views and Estimates Report' on the president's FY 2005 budget. Whatever support might have been building among democrats in the House to support the initiative was now taking a nose-dive. The report said, "Until the Congress has better information on which to judge the long-term cost of the president's Moon/Mars initiative, NASA's FY 2005 funding request should be reallocated in a manner that strengthens NASA's existing programs, helps address the backlog of deferred maintenance at NASA's facilities, ensures that the Shuttle will continue to fly safely for as long as it is needed, ensures that the International Space Station will be a safe and productive facility, makes a start on a replacement means of getting U.S. astronauts into space, and enables the analyses that will be needed to develop a viable and sustainable exploration agenda.

The next day the House Science Committee held a hearing titled 'Perspectives on the President's Vision for Space Exploration'. House Science Committee Chair Sherwood Boehlert (R-NY), known to be generally sympathetic to Sean O'Keefe, was blunt as he opened the hearing. He said, "I think all I need to say about my views this morning is to reiterate that I remain undecided about whether and how to undertake the exploration program. I would add that, as the outlines of the likely fiscal 2005 budget become clearer, my questions about the initiative only become more pressing."

Ranking Minority Member Bart Gordon (D-TN) stated, "I support the goal of exploring our solar system. However, until I am convinced that the president's plan to achieve that goal is credible and responsible, I am not prepared to give that plan my support."

Norman Augustine, renowned observer of the aerospace industry, and chair of the Augustine Commission which reviewed the previous President Bush's space exploration initiative, was blunt about the sort of consensus that needed to be constructed to support the current president's space vision, "If America is to have a robust space program it is critical that we build a national consensus as to what that program should comprise. If, for example, we are to pursue an objective that requires twenty years to achieve, that then implies we must have the sustained support of five consecutive presidential administrations, ten consecutive Congresses and twenty consecutive federal budgets—a feat the difficulty of which seems to eclipse any technological challenge space exploration may engender.

Responding to Augustine's prepared statement Rep. Nick Lampson (D-TX) said, "Mr. Augustine states in his written testimony that 'it would be a grave mistake to try to pursue a space program on the cheap. To do so is in my opinion an invitation to disaster.' I could not agree more."

The other witnesses (Donna Shirley, Lawrence Young, and Lennard Fisk, and Michael Griffin) all evidenced enthusiasm for the prospect of a new program of exploration, but all cautioned, in one way or another, that more thought needed to be given to the actual plan, the costs, the final goals, and the overall implications to existing programs if such a new plan were adopted.

No one in the room stood up and pledged their unqualified support for the president's plan. Indeed, those in a position to advance it, Boehlert and Gordon, made it very clear that they needed much more detail before they made any move to openly support the plan.

Congress had clamored for a new space vision in the aftermath of the Columbia's destruction—one led by the White House. Now they had one—endorsed by the president. Either NASA had not done their homework, or Congress did not know just what it was looking for. It would soon become clear that both factors were in play.

The Senate Comes Through

The specter of a detached White House rose again on 12 March 2004. Sen. Bill Nelson commented on the Senate floor about a late night vote wherein the Senate voted on an amendment (S.AMDT.2733 to: S.CON.RES.95 – the Senate Budget

Resolution) "whose purpose was to provide full funding for NASA's FY 2005 space exploration initiatives."

The amendment had the finger prints of the space states written all over it and was a vote of full support for what the president had proposed. Sen. Sessions (R-AL) had introduced it (after conferring with Sen. Brownback (R-KS) and Sen. Cornyn (R-TX), Sen. Nelson (D-FL), Sen. Shelby (R-AL), Sen. Graham (D-FL) and Sen. Hutchinson (R-TX) had all co-sponsored it. The amendment passed by a voice vote.

Nelson complained that he felt the White House had been missing in action in lobbying to generate the votes needed for its passage. Nelson continued, "My pleas in the course of our deliberations in the Budget Committee to get the White House to step forward and to support its request for its full funding at a level of $16.2 billion, went unheeded... It was only when Senator Sessions and Senator Shelby each put their foot down to let the chairman of the Budget Committee know that their votes on final passage were questionable unless that was brought up to the level of the president's request did we successfully get inserted into the budget an amendment that would bring NASA up to the $16.2 billion."

In frustration Nelson then said, "I call on the White House. I call on the leadership of NASA. We cannot take for granted just because the president has announced a major new initiative that it is going to get funded. Indeed, we are swimming upstream. The immediate reaction of the American people to the president's initiative was they didn't support it. There is only one person who can lead the space program. That is the president or the vice president. A Senator can't lead it. The administrator of NASA can't lead it, particularly on bold new initiatives. It has to be the White House that leads it."

As it happens this whole process was not as out of control as it might have seemed to Nelson at the time he spoke. There were hands, guided by the White House, working behind the scenes for more than a week to help orchestrate this vote. The fact that both senators, from both parties, from Alabama, Florida, and Texas were in on the initial sponsorship, and the fact that Brownback had lent his support to Sen. Sessions and Sen. Nickles. The vote came after all of the rhetoric had been expended, and happened in a fashion such that Nelson could claim some leadership in marshalling the amendment through. "Victory has a thousand fathers," recalled O'Keefe when asked about this vote and the process that led up to it.

House Budget prospects

While the Senate was being generally supportive, the House was not, at least not initially.

In late March, Sean O'Keefe, and several senior aides met with the so-called Blue Dog Democrats, a group of conservative House members that included Charles Stenholm of Texas, Mike McIntyre of North Carolina, Robert 'Bud' Cramer of Alabama and Gene Taylor of Mississippi, as well as other Democratic budget hawks.

The group vented their concerns about NASA's budget and got in return what one attendee called a detailed review of how the space agency plans to pay for the new space effort. The result was the Blue Dogs agreed to support an amendment to the House version of the budget resolution granting the full NASA request.

According to congressional sources, several House members complained Bush had failed to say anything more about the Moon-Mars plan since his 14 January speech, and his silence had been interpreted as a cooling of support. The group was told the White House was silent, not because Bush was rethinking his grand space plan, but was instead trying to avoid further politicization.

One source said that Bush would 'keep his powder dry until the myths, legends, and political barbs on this strategy subside,' and the president probably would speak again about his space plan sometime late in his re-election campaign.

NASA also received new support from the Republican Study Committee, another group of fiscal hawks. Reps. John A. Culberson of Texas, and Tom Feeney and Dave Weldon, both of Florida, as well as Roscoe Bartlett of Maryland, helped gain the group's support for NASA, a source said.

Although all these recent signs were positive, the space plan still faced an uphill battle. The Blue Dog amendment was far from a floor vote, and the Senate Appropriations subcommittee that controls NASA funding had yet to weigh in on its recommended final amount for FY 2005. Considerable support remained for freezing NASA's budget at 2004 levels.

Boehlert spells it out

On 21 April 2004 Rep. Boehlert addressed a session held on Capitol Hill under the auspices of the American Institute of Aeronautics an Astronautics (AIAA). While still expressing broad enthusiasm for the general notion of a new space policy, Boehlert hammered away at the lack of details he saw in the president's outline for what he wanted NASA to do.

As for how he saw the prospect of NASA getting the funds in FY 2005 that the president had requested, Boehlert's predictions were not good news, "As part of the exploration initiative, the president has proposed increasing the NASA budget by 5.6

percent in the next fiscal year, to about $16.2 billion. I just can't imagine that that's going to happen, and I don't think it should. Total federal non-security, domestic discretionary spending in fiscal 2005 is likely to increase by less than half a percent. Congress may even freeze spending, as the House voted to do in its Budget Resolution. In such a budget, should NASA receive almost a 6 percent increase? Is it the highest domestic spending priority? I don't think so, and I doubt my colleagues will either."

Boehlert predicted that, "Congress isn't likely to even take up the NASA spending bill until after Election Day. (I'm not proud of that, but its reality.) That means that for at least a month, and potentially for several months, NASA will be funded by a continuing resolution. That, in turn, means that for some portion of next year, NASA will be flat-funded and will not be allowed to start new initiatives. That alone could delay aspects of the exploration initiative."

On that same day Sean O'Keefe sat before a hearing held by the House Appropriations Committee's Veterans Affairs, Housing and Urban Development and independent agencies subcommittee. The Subcommittee's chair, Rep. Walsh (R-NY) was as blunt as his colleagues in the house science committee saying, "I cannot commit this Congress and future Congresses to a [space] program that is undefined." Echoing Young's stance, ranking minority Member Rep. Alan Mollohan (R-WV) said much the same thing.

As was the case in all other such situations. Sean O'Keefe replied that he stood ready to provide as much detail as was needed in order to allay the concerns of the committee. As April rolled into May things did not look as bright in the House as they did in the Senate.

The House Comes through

Nonetheless, things did eventually turn NASA's way—just barely. On 19 May 2004 the House had a chance to vote on S.Con.Res.95. By a close vote of 216 – 213 they agreed to the conference report. The vote breakdown was (Republicans) 216 for, 9 against, 3 not voting. On the Democratic side the vote was 203 against (plus 1 Independent against) and 2 not voting. This was quite clearly a vote along party lines, and clearly not a vote only on the issue of the president's space initiative.

However, among those voting against the Bill was Rep. Gordon (R-TN), ranking minority member of the House Science Committee; Rep. Mollohan, ranking minority member on the House Appropriations Committee's Veterans Affairs, Housing and Urban Development and independent agencies subcommittee, Sheila Jackson-Lee (D-TX) an ardent supporter of space topics in Texas, Rep. Bud Cramer (D-AL), and

Rep. Lampson (D-TX) who had introduced a bill calling for a robust space exploration vision in the 108th and 107th congresses.

H. Rept. 108-498, which was the product of this vote, said, "The conferees support the president's Vision for Exploration and believe the fiscal year 2005 funding for Function 250 should provide sufficient funding to initiate the process. Additionally, the bulk of the requested increase for fiscal year 2005 is for return to flight of the space shuttle and continued assembly and operations for the International Space Station. The Conferees hope that these two must-fund requirements will be taken into account during their consideration of the NASA appropriation."

While the forces had aligned to bring the president's FY 2005 request for space ahead, more or less intact, the prospect that it would be approved by House-Senate appropriations committees, and that this would happen before the 2004 election was increasingly remote.

Moreover, the Conference agreement does not necessarily guarantee that NASA gets all the money it needs for the new exploration initiative. Indeed, it may get a rather small portion and be tied to FY 2004 levels.

Still a long road ahead

Should Congress wind up funding NASA at last year's level, NASA would need to slash spending on a variety of programs to keep new activities running. Of the $866 million increase sought in the FY 2005 budget, 85 percent would be needed to cover repair costs on the shuttles and the space station assembly. $374 million would be needed for the shuttle repairs, $365 million for the ISS leaving only $136 million to begin the new space exploration plan.

In an account that includes science and space programs—such as NASA's spaceflight projects and other projects within the Energy Department and the National Science Foundation—the conference committee approved $23.8 billion in spending for FY 2005. Normally, some two-thirds of that total would go to space programs. The actual amount NASA will receive, however, must be determined by the appropriations committees of both houses of Congress, neither of which had yet taken up NASA's spending request.

A reauthorization bill for the civil space agency was expected to be finalized and head for a House floor vote before Congress adjourns for its summer recess. Even if that occurred, though, a final budget bill for NASA was unlikely to emerge before the new fiscal year began. Instead, Congress would probably approve a resolution that allows the federal government to continue spending money, usually at the previous year's level, until legislators authorize the new budget. Following what has become something of a routine over the past decade or so, continuing resolutions could fill the budget gap well into FY 2005. In fact these stopgap spending measures may be needed through the entire fiscal year.

One reason for the delay in congressional action on NASA's budget is the shortened work calendar for 2004 because of the political campaigns. The House and Senate space committees also have been reluctant to act on the administration's space request, citing the need for additional information and details on the project.

Meanwhile, though NASA's budget may be approved by the House in July, Senate action before the August recess, and during the abbreviated fall session, seems unlikely. Congress is set to adjourn in October so House members and some senators can campaign for re-election, as well for their presidential standard-bearers.

As such, whatever the final fate for the FY 2005 kick-off funding request for President Bush's new space vision, this decision may not be made until after the next occupant of the White House is chosen.

Delay Takes Over

Speaking on 21 April Boehlert moved to the topic of a NASA Authorization Bill he was working on. Not having managed to move an authorization bill out of the committee in the last congress, an attempt was underway to make one happen in this congress. Boehlert said, "...we ought to have a NASA reauthorization bill that will lay out the broad blueprint of how the human space flight program should move ahead. An authorization bill will enable Congress to have the broad debate you call for to (quote) 'come to a clear agreement on the goals of the nation's civil space program.' ... Coming up with such a bill and moving it through the Congress will not be easy, but we are committed to pursuing it. I hope we will be able to introduce a bill around July 4th and move it through the House in September."

As it would happen, Boehlert would not make much headway with his authorization act. Within a few weeks, House Majority Whip Tom DeLay (D-TX), whose district included NASA JSC, took over responsibility for the development of authorizing legislation after being frustrated with the lack of progress and skepticism amongst committee republicans in the president's policy.

In May 2004 DeLay would address the 'National Keep It Sold' group, an advocacy organization of various business community interests in states with large space spending. DeLay made no secret of his enthusiasm for the plan—titling his remarks 'On to Mars'. The remarks were short on substantive detail but he made it very clear that he was in the president's corner, quite a contrast to what he said in the past, nearly two years to the day, when he told Sean O'Keefe, "You have a timid and anemic plan for human spaceflight."

Outside Help – Wanted and Unwanted

Large advertisements for aerospace products and systems are a common feature in newspapers read inside the beltway in Washington. Despite the cost of such ads, and the readership of local newspapers, the target audience is astonishingly small: members of Congress and their staff. But given the financial consequences involved in such programs, tossing thousands of dollars at the Washington Post to keep the issue visible is a trivial expense by comparison.

Direct public lobbying by an agency for its programs is not considered acceptable in Washington; neither is an overtly orchestrated effort by proxies. But the use of proxies (industry associations or company lobbyists) who push for things in a variety of slightly veiled, as well as subtle modes, is standard fare. Indeed, often times those pushing for one program or another don't need an agency to tell them what is going to result in a sale for their company. Sometimes the tactics adopted by the private sector and industry groups synch with what an agency is trying to accomplish. More often than not the synchronization is not perfect, with industry pushing one aspect of a program or policy while the agency is hoping things will head in a somewhat different direction.

And then sometimes nothing is in synch and everyone is confused. Such is what happened when the new space policy was rolled out.

As soon as the outline of the new space policy became clear, Washington-based industry representatives and lobbyists began to meet informally. It was abundantly clear to all in attendance that a rising tide lifts all boats and that coordinated efforts, if done properly, would be a benefit to all involved. Nonetheless, with no formal entity or organization to act as a coordinating body or clearinghouse, efforts to create one began. To be certain there was no shortage of industry groups and organizations, but the broad nature of what many saw as being needed seemed to call for a new entity specifically designed to coordinate efforts relating to the new space policy.

At first, the two aerospace giants, Boeing and Lockheed Martin pursued separate paths. Boeing tended to favor some initial efforts being suggested by the Space Foundation, efforts which seemed to have some favorable blessings from NASA. Lockheed Martin along with Northrop Grumman, Alliant Tech Systems, and Kistler drifted towards an effort being drawn together by DFI. United Space Alliance, a joint effort between Boeing and Lockheed Martin hedged its bets and waited to see which effort would gather momentum and receive favor from NASA.

Four major organized efforts began to emerge in the weeks following the president's speech. Only three would survive the formation process. The first to begin organizing was the 'Coalition for Space Exploration' which was spearheaded by the Space Foundation. The second, which began to form shortly thereafter, was 'Space Exploration Partners' which was being pushed by DFI International. Then there was a group formed by space state business and political leaders called 'National Keep It Sold' – an odd name which was replaced in May 2004 with the more logical 'Citizens for Space Exploration.' Lastly there was the 'Space Exploration Alliance' – a joint effort by most of the nation's large space advocacy and professional organizations backed by Walter Kistler.

Space Exploration Partners was initiated by DFI International, a Washington DC-based consulting firm. The prime movers in this effort were Royce Dalby and Lori Garver. Garver, well-known in space circles had most recently tried to mount a PR campaign to become 'AstroMom'. This would have involved flying in a Russian Soyuz to the International Space Station. After a flurry of PR, and some seed money from DFI her effort faded when money did not materialize . N'Sync singer Lance Bass's bid to fly on the same Soyuz soon eclipsed her attempt. They joined forces to some extent, but Bass' effort soon fizzled when no one would cough up the money to pay for his ride.

Previously, Garver had been Associate Administrator for Policy and Plans at NASA (a Clinton political appointee) and before that the Executive Director of the National Space Society. Garver was someone with space exploration clearly evident in her blood, even having her gall bladder removed so she could qualify for the ride into space.

Space Exploration Partners was described in a briefing prepared in mid-February 2004 to prospective members as being "an industry coalition organized around the new NASA initiative." Among the prime activities of Space Exploration Partners would be to "promote and track the progress of the president's initiative; coordinate messages among the interested parties; develop public outreach strategies; and act as a communications conduit to and from Congress and the Administration."

The approach to be used was to have "regular meetings with congressional staffers to access receptivity to coalition concerns; writing and coordination of letters and op-eds in newspapers and critical districts; coordination with industry and public advocacy groups to develop a cohesive message; outreach to peripheral industries (e.g. mining, construction) to broaden industry support; and outreach to the science community to build a stronger base."

The core purpose of the effort was very simple. Again, according to the DFI presentation, "The key is the bottom line – more money and support for NASA, in pursuit of a sustainable program of exploration, will mean more business for the members of the Space Exploration Partners." DFI had thought to charge its members $5,000 a month to finance the effort—which would be led by Lori Garver.

There was one small problem with what this group sought to do: it sought to be the coordinating body for NASA and the Bush Administration. But neither NASA or the White House wanted this sort of help.

Garver said she had received indications from NASA and the White House that they were receptive to this idea—given that NASA could not go out and openly lobby or market the new space policy—and she told meeting attendees that NASA was indeed favorably disposed to their effort. This was echoed in an email sent out to invite people to attend a meeting by DFI's Royce Dalby which said, "DFI has been encouraged in forming this coalition by numerous NASA and White House officials, who believe that a well organized coalition of industry that focuses on all aspects of marketing and tracking the initiative will ensure that the vision has the greatest possible success." As such people paid attention to DFI's efforts thinking that they had been blessed by the White House and NASA.

Two meetings were organized by DFI. Brett Alexander from the White House attended the first meeting. John Schumacher, NASA's Chief of staff, and Lee Forsgren, NASA's AA for legislative affairs were also scheduled to appear at a DFI meeting but NASA started to distance itself from the DFI program.

Mixed signals reverberated around Washington as to whether NASA had indeed blessed this effort or whether it held the plan in disfavor. After several weeks of fumbling, NASA decided that it would not support the Space Exploration Partners. Neither DFI nor Garver were at fault for the mix-up. It was NASA's own confusion as to which industry association or nonprofit it would select to take the public relations lead that lies at the center of the issue.

As this was all happening the Space Foundation was working on establishing the Coalition for Space Exploration; one which sought to harness the resources of the aerospace industry to promote a broader public awareness of the new space policy. Self described as being 'a collaborative effort whose mission is to ensure the United States will remain a leader in space, science and technology – key factors that will benefit the nation's economy, gratify our need to explore and maintain our national security', the coalition, unlike DFI's Space Exploration Partners did not seek to act as an intermediary of sorts between congress or the White House. Rather they sought to direct their efforts to the general public at large.

Membership of the Coalition would eventually grow to include Alliant Techsystems, Boeing, Lockheed Martin, United Space Alliance, Space Foundation, National Space and Satellite Alliance, Aerojet, Analytical Graphics, California Space Authority, Florida Space Authority, Hamilton Sundstrand, Honeywell, Moog, Pratt & Whitney, Northrop Grumman, and Harris. Members paid fees that were used to fund the purchase of advertising space in prominent national newspapers and to pay space.com for an online petition where supporters of the new space policy could sign up to express their support.

On 10 March 2004 NASA arranged a get-together of members of Congress and their staff at NASA HQ. Members were not shy in giving Sean O'Keefe their view of where the president's space policy was going and that O'Keefe needed to step up his efforts to have a visible and more effective presence on the Hill. O'Keefe replied to the effect that he 'got the message' and was already working on some enhancements in his legislative affairs operation.

Some members said that it would be helpful to get the industry organized. O'Keefe spoke in less than positive terms about an effort by some industry consultants to form a coalition of sorts which sought to speak on NASA's behalf. O'Keefe said NASA was not endorsing any consultant group-led effort. He was referring to the DFI-led effort which had dissolved several weeks earlier.

As the Coalition for Space Exploration was forming and Space Exploration Partners was fading, Kistler Aerospace Corporation (currently attempting to emerge from bankruptcy) and the Foundation for the Future (funded by Walter Kistler) invited a dozen or so space industry and policy leaders for an all-expenses-paid seminar on 2-3 April in Seattle to plan public outreach strategies for the president's new space initiative. The heads of all space advocacy groups were on the invitation list as well as former NASA space policy officials. The event was held immediately after the Space Foundation's annual symposium.

At this meeting a schism of sorts formed. The Space Foundation was wary about creating another entity with a name—given that they were now forming their own 'named' effort. Other participants wanted to have a discrete entity for everyone to rally around. A subsequent meeting—again with expenses paid for all out of town participants by Kistler—was held in Washington DC. Out of this effort emerged the 'Space Exploration Alliance' whose members included the Aerospace Industries Association, Aerospace States Association, American Astronautical Society, American Institute of Aeronautics and Astronautics, California Space Authority, Florida Space Authority, The Mars Society, National Coalition of Spaceport States, the National Space Society, The Planetary Society, ProSpace, the Space Access Society and the Space Frontier Foundation.

In a press release issued with the announcement of the Alliance's formation the Alliance stated that their first goal as a group was "to work for broad Congressional support of the new national vision for space exploration outside of low Earth orbit, which they refer to as Moon, Mars and Beyond." The Alliance would do this by working "to secure first year funding for the initiative, which they view as a necessary first step for in-depth planning of the exploration program to commence in earnest."

Unlike the Coalition for Space Exploration which charged its members fees on the order of $10,000, the Space Exploration Alliance was a much less structured organization, with each of its members retaining much of their own agendas and focusing on some common, core points to promote. A few weeks later they announced that they would sponsor a 'Moon-Mars Blitz' on Capitol Hill in July 2004. Modeled after ProSpace's small, but enthusiastic 'March Storm' events, the organizers hoped to have a large number of space advocates descend on Capitol Hill for a citizen's push for the new space policy.

While these events were underway, a fourth effort was becoming re-organized (having initially come into existence in 2002). Initially known as 'National Keep it Sold' (as-in keep the notion of space exploration sold) they soon changed their name to 'Citizens for Space Exploration'. Described as "a grass-roots effort made up of small business owners, educators, students, business leaders, elected officials and other citizens in Texas, Florida, Alabama, Mississippi and Louisiana. Citizens for Space Exploration works year round to boost public and legislative support for America's human space program, particularly the Space Shuttle and International Space Station (ISS) programs."

Not everyone was happy with the new space policy. One of the more curious examples came from the AFL-CIO. On a special website devoted to taking on the Bush Administration the Union said, "On Jan. 14, President George W. Bush announced a plan to put people back on the Moon and send astronauts to Mars. Meanwhile, every day in America: 85,444 workers lose their jobs, 14.7 million workers are jobless, underemployed or have given up looking for work." etc. On the right side of the page, three photos show average Americans with quotes such as, "We need jobs." "We've got to protect Medicare." and "$100 billion to send a man to Mars?" A link allowed visitors to view a TV ad which followed along similar themes. Given that the overwhelming majority of any increase in space spending would create jobs, many of them union, caused many to wonder what the real motivation was behind the website and whether anyone had done a sanity check before letting it and the TV spot out.

The Rollout Plan (or lack thereof)

So, is NASA at fault for not having a comprehensive rollout plan ready to go the moment the policy was announced? In a perfect world, yes, NASA could have done better. But this was Washington DC. Things leak. People spin leaks. Spun leaks beget false rumors and eventually you can spend more time shooting down false rumors than focusing on the actual policy development process itself.

NASA could have begun to set up an outreach and rollout plan for the new space policy well ahead of the eventual rollout date. However NASA would not have been able to progress too far before the small group of policy architects would have had to share more information than O'Keefe or the White House would have liked to reveal.

Moreover, even if NASA managed to keep the lid on things, given the fluidity with which things tended to change, it would have been an outreach effort whose core fundamentals would have been constantly in flux up to almost the last minute.

Lastly, had NASA begun to work on a rollout plan too far in advance it would have invariably got back to the White House. It might well have made NASA seem overly presumptuous—even over confident with a certain outcome at a time when the final details, even the decision whether or not to do anything at all, was still up in the air with the final call resting with the president. Some of the major points of the policy were still being nailed down as a decision package was being prepared for the president in mid-December. Over confidence on NASA's part might have affected the final White House decision process itself.

Add in the fact that the story almost got out several times in December, and at least once in January, before UPI broke it a week ahead of the time when the White House would have preferred it to be released, and it becomes clear that NASA did not have that much time to lavish upon coordinating a multifaceted roll out plan.

Nonetheless, once the policy was known, the agency could have put together a tighter package and a rapid response team to react to roadblocks and unexpected events as they presented themselves. However, once everyone became familiar with what the policy did and did not contain, what the agency did and did not know, and where NASA and the White House were willing to go in order to move everything forward, support did begin to solidify and move in the direction of support for the policy.

If you think that was hard

Selling the plan was hard enough. Making the hard choices that went with it would be even more difficult. NASA had grown rather fond of the space shuttle and the International Space Station. Now one would be retired at the end of the current decade and the other would have served its purpose before the end of the next.

NASA wasn't ready to let go of either.

*   *   *   *   *

10

HARD CHOICES

Lighting out for the (new) territory

Poised on the cusp of a new century, and the aftermath of a second shuttle accident, people started to ask why we were doing things in space. Was the plan simply to circle Earth, but not go anywhere other than round and round? Or were we going to try and do what we once did, head for the Moon and then go beyond and resume the human exploration of our solar system?

A cursory glance at the Bush space policy would suggest, at least in terms of what Bush would support, that the phrase 'humanity is headed out into the cosmos' is a pretty firm indication that it was time to break out of low Earth orbit once again and actually go somewhere.

But the NASA that received this new challenge would have to do so with limited resources; much of which had to be adjusted from those at hand. Among the resources needing modification were the space shuttle fleet and the International Space Station (ISS). Once returned to flight status, the space shuttle would now be flown only until the ISS was completed. The shuttle would then be retired and Americans would ride into space aboard a new, modular spacecraft, the Crew Exploration Vehicle (CEV). As for the ISS, the array of research projects to be done by the U.S. would be refocused, culled, and redirected toward the flight certification of humans for long duration travel to the Moon—and then on to Mars and other locations.

Both decisions would lead to much consternation and would challenge constituencies who did not necessarily want things to change—at least not as fast as Bush's new plan would require.

Many saw the shuttle, the flawed and sometimes lethal ride into space, as nothing short of a national treasure—if not a technological marvel despite its limitations. The country had invested vast resources to develop and maintain the fleet; even replacing an orbiter when Challenger was lost. Some felt the fleet could easily fly for several more decades, not unlike the venerable B-52, whose aircraft were significantly older than its pilots. Others felt that this system was at the end of its useful life. Were it not for the fact that the design of remaining ISS elements depends on shuttles to be launched and the fact that there was no indigenous replacement coming online, the shuttle fleet might have faced immediate retirement.

Shuttle & ISS

The compromise that was struck was to retire the shuttle as soon as all the components were in place to complete the space station's assembly. Meanwhile, the CEV would be developed to allow U.S. access to space. But the CEV might never even visit the ISS since the vehicle is planned to come on-line around 2014. Around that time, according to the new policy, the U.S. would be implementing the other difficult decision: scaling back its use of the ISS and looking outward to the Moon and Mars.

While many previous notions for a large, permanent U.S. space station were discussed over the years, the one that finally caught on, first known as 'Space Station Freedom', and later as the 'International Space Station', was always touted as providing all things to all interests. Therein lay its greatest utility as well as its greatest vulnerability. Since such a vehicle could not cater to everyone—but was forced to try. In the process, many potential users suffered.

Moreover, the space station was never able—at least from the U.S. perspective—to have a simple raison d'être—a singular, core mission you could put on one sheet of paper. As such, what was being done on the ISS was diffuse, hard to explain, and even harder for people to get excited about. Add in the fact that the project was plagued with cost overruns and delays from its very onset, any discussion of its value was often eclipsed by annual examples of its mismanagement.

Let's fly the Space Shuttle forever.

The space shuttle arose from design concepts formulated in the early- to mid 1970's. It first flew, several years late, in 1981. Up until the Columbia accident NASA was at work on efforts to upgrade the shuttle fleet such that it could fly until 2020—and perhaps even beyond. It was possible that the shuttle fleet would fly for an entire human generation.

Some proponents of the notion of flying the shuttle fleet for several more decades would often point to the B-52 fleet. Many of the huge bombers have been in constant service since the early 1950's and were held as an example of what a properly maintained airframe and components could do—after all each shuttle's airframe was designed to be capable of flying 100 missions. But there is a vast difference between a truly operational military vehicle designed to be maintained for multiple, frequent sorties, and a spacecraft that flies a few times a year and needs a major overhaul between each and every flight.

In a letter dated 25 March 2002 titled 'Reassessing Space Shuttle Upgrades Strategy' Frederick D. Gregory, then the Associate Administrator of NASA's Office of Space Flight asked the Johnson Space Center (JSC) director to "develop a strategy as part of initial planning efforts to identify upgrades and supportability investment that may be required to maintain the space shuttle fleet capability to fly safely through 2020." Gregory noted that, "No commitment or decision has been made by Headquarters regarding shuttle activities beyond 2012."

NASA had been engaged for a number of years in a series of safety and reliability upgrades to the shuttle which had been brought together under the title 'Service Life Extension Program' (SLEP)—a process derived from the way in which military fleets were kept in service. To be certain, what had been contemplated within the confines of SLEP would cover some of what would be required to fly the shuttle fleet for extended mission times in space in the near term. But clearly, much more would be needed if the Shuttle were to fly for another two decades instead of one.

In hearings held on 18 April 2002 before the House Subcommittee on Space and Aeronautics, Gregory "tasked the Space Shuttle Program to assess upgrade investments required to safely fly the Space Shuttle through FY 2020. This does not in any way indicate an agency decision to fly Shuttle until 2020; however, we believe it is prudent to understand this information as a contingency."

The report prepared in response to this request, entitled 'Space Shuttle Program 2020 Assessment' was finished and presented to Gregory by Acting Deputy Associate Administrator of Space Shuttle, Parker Counts on 21 August 2002. NASA had planned to continue with analysis activities through August of 2003 with specific recommendations offered for new studies that would start in FY 2004.

Throughout the second half of 2002 and the first half of 2003, both Mike Kostelnick (Deputy Associate Administrator for International Space Station and Space Shuttle Programs) and Gregory made pledges to the media to release the report. Despite these pledges it was not until May 2003, and multiple requests by the media, that it was finally released.

The report noted that "the beginnings of erosion are evident in many areas common to multiple shuttle elements; flight subsystems, ground facilities, special test equipment, vendor support, process escapes, and human capital. Strategy must address all these areas." The report showed a cost (adjusted for inflation) out to 2020 of $8.67 billion as being necessary to deal with these issues. This amount was one order of magnitude of what was required to build the shuttle fleet in the first place. Many wondered if this was about to become a clear case of throwing good money after bad. Others had long ago arrived at that conclusion.

Dashed Expectations

By the time NASA finally got around to releasing this report, the Columbia accident had already thrown the issue of flying the shuttle for such a long period into doubt. Indeed, when the CAIB issued its final report, that prospect became even less likely— indeed, remote.

The CAIB stated, "It is the view of the Board that the present Shuttle is not inherently unsafe. However, the observations and recommendations in this report are needed to make the vehicle safe enough to operate in the coming years." The CAIB recommended that "prior to operating the Shuttle beyond 2010, develop and conduct a vehicle recertification at the material, component, subsystem, and system levels. Recertification requirements should be included in the Service Life Extension Program (SLEP)."

Given that Sean O'Keefe had pledged that the agency would adhere to all recommendations made by the CAIB, and that current plans showed that the U.S. would still be utilizing the ISS in 2010, NASA was looking at an almost certain recertification of the shuttle. To confound things, no one quite knew what it actually meant to 'recertify' the shuttle—except that it would be very expensive.

In the coming months NASA would issue a regular series of updates to its 'Implementation Plan' for both returning the shuttle fleet to flight and how this would affect ISS program activities. It became abundantly clear that this task was not going to be as easy as some expected, with several projected launch dates for the next shuttle mission, STS-114, being slipped in quick succession.

In late March 2003, a matter of weeks after the accident, NASA was working towards launch dates for the next shuttle mission of no earlier than 21 July 2003. This target launch date was then slipped to no earlier than 1 October 2003 in early April. That date soon evaporated as it became clear that nearly another year would be needed. In October 2003 the projected launch date was moved from September to October 2004. In February 2004 NASA moved the date to March 2005. In one year, a 5 month delay grew into a 24 month delay.

As NASA moved toward the announcement of the new space policy, its only means of carrying humans to orbit had been revealed to be far more flawed and antiquated than its fans would have thought just a year before. Whereas perfectly rational managers at NASA saw the prospect of flying the shuttle until 2020 as not only doable, but desirable, it now became clear that every launch would have to be fought for and that the shuttle's days—at least as a carrier of human cargo—were numbered.

When the president spoke on 14 January 2004 the ISS and space shuttle programs, which were often viewed (and operated) as if they'd never come to an end, were given endpoints—as far as the U.S. was concerned.

President Bush said, "Our first goal is to complete the International Space Station by 2010. We will finish what we have started, we will meet our obligations to our 15 international partners on this project. We will focus our future research aboard the station on the long-term effects of space travel on human biology."

Although Bush did not enunciate a time frame, briefings by NASA would later show that the U.S. would operate this new focused research program until such time as it got the answers it would need to send humans on long duration missions. At that point, the U.S. would begin to reduce its participation in the ISS program (although to what level this would be was unknown). With lunar landings as soon as 2015, one would expect that ISS usage for research would start to wind down shortly thereafter.

Bush went on to say, "The shuttle's chief purpose over the next several years will be to help finish assembly of the International Space Station. In 2010, the space shuttle—after nearly 30 years of duty—will be retired from service." As such there would likely be no need for recertification if the retirement date was close to this target. Moreover, the upgrades needed to fly until 2020 would be reduced to those needed to fly until 2010.

In addition, according to Bush, the shuttle would be replaced—not augmented: "Our second goal is to develop and test a new spacecraft, the Crew Exploration Vehicle, by 2008, and to conduct the first manned mission no later than 2014. The Crew Exploration Vehicle will be capable of ferrying astronauts and scientists to the space station after the shuttle is retired." Bush completed the core aspects of the new space policy by saying, "Our third goal is to return to the moon by 2020, as the launching point for missions beyond."

Of course this plan would lead a gap between 2010 when the shuttle would retire and 2014 when a CEV would be available to carry humans. NASA's reply was that the U.S. would depend on Soyuz spacecraft to carry Americans to the ISS. NASA Comptroller Steve Isakowitz told reporters several days later after the president's announcement that the U.S. may begin to ramp down the use of the ISS around the time that the CEV begins to carry humans, such that trips to the ISS by the CEV might not be necessary. He said that it was possible that the CEV would never visit the ISS. Given the high orbital inclination of the ISS it is somewhat unlikely that missions to the Moon would be mounted from there. However some preliminary test flights—including a test of docking systems, might be done with the ISS as a destination.

Not so fast

Discussions about what to do with the shuttle fleet were not new. After the Challenger accident, NASA was directed by the White House to put an end to flying overtly commercial payloads on the shuttle fleet. In addition, NASA was admonished by the Rogers Commission that the shuttle fleet was not 'operational' (despite President Reagan's proclamation to that effect after the STS-4 landing), nor was it likely to ever be 'operational' in the sense that NASA had been attempting to convey.

In subsequent years, NASA would flirt with the possibility of commercializing one aspect or another of shuttle operations. In creating the SFOC—the Shuttle Flight Operations Contract—which was 'won' by the only bidder United Space Alliance, a joint partnership of Rockwell International and Lockheed, NASA did manage to move much of what had been government accomplished tasks into the private sector, but was stymied by the Goldin administration from going much further.

Regardless of how NASA sought to alter the flavor of how it operated the shuttle fleet, the notion that it would actually do away with the fleet was considered heresy. Even when shuttle successors such as the Orbital Space Plane were proposed, the date at which the shuttle would actually stop flying was so far in the future as to be moot. As such, it was unlikely, prior to the president's speech, that the shuttle system would necessarily go away entirely.

Using shuttle technology in new ways

One thing NASA and its contractor community loves to do is 'evolve' systems. That is, take an existing design and enhance its capabilities or give it a new function. The idea is that since you have already invested money into the design you will save money by using it as the departure point for new systems. Sometimes this is true. More often it is not. The idea of creating Shuttle-Derived Launch Vehicles (SDLV) is not new. Even before the first flight of Columbia in 1981, Rockwell International and other contractors talked about enhancing the capacity of the shuttle, and improving its reliability by replacing its Solid Rocket Boosters with reusable liquid boosters. Other concepts involved stretching the shuttle orbiter or replacing the orbiter itself with a large cargo canister with engines placed underneath roughly where they'd be on a shuttle orbiter.

During NASA JSC's flirtation with a 'Mars Reference Mission' architecture in the late 1990s, a NASA MSFC concept called the Magnum booster was promoted by George Abbey. This system would take the existing shuttle system and progressively augment it with Liquid Fly-Back Boosters (LFBB) and later replace the shuttle orbiter with stages capable of carrying as much as 80 metric tons into low Earth orbit. A series of launches would be used to assemble a spacecraft capable of going to Mars and back. This planning effort eventually shut down when the Clinton White House became annoyed with NASA's continued insistence on promoting human missions to Mars.

Another shuttle derived concept was the Shuttle-C (C for 'cargo') developed by NASA MSFC. NASA even went so far as to use some leftover engine testing hardware to construct a full scale replica in an unused hangar in Alabama. After it became clear that NASA had no use for this concept it had put forth, the mockup sat around gathering dust. After serving for years as a popular stop for NASA employees visiting Marshall, the mockup was disassembled and sent to the scrap heap.

And yet another incarnation occurred during the Space Station Freedom redesign in 1992-1993 when an all-in-one alternate space station design to be launched using shuttle technology was suggested by the 'Option C' team which was based at JSC. This concept never made it past the PowerPoint stage.

NASA has continued to fiddle with various shuttle derived concepts ever since. At the February 2004 SLEP Summit, where all of the companies involved in Shuttle operations meet, on page 11 of his presentation, Michael Kostelnick, showed a chart titled 'Supporting Exploration Vision'. This chart referred to the fact that "shuttle-derived requirements [are] still evolving" and that the "shuttle infrastructures provide unique exploration enablers." On that same chart is a picture of a shuttle-derived heavy-lift vehicle lifting off while other variants are shown—including configurations where the cargo is attached like the shuttle is and one with the cargo directly inline with the external tank. In addition, one configuration is depicted which shows a Crew Exploration Vehicle (with launch escape tower) atop a shuttle Solid Rocket Booster.

As such, NASA is very clearly thinking about how to keep familiar shuttle technology in play for decades to come—even if the current space shuttle orbiter fleet is not carrying humans. One option also discussed is to fly the current space shuttle orbiter unmanned. While the intent was always to have the shuttle vehicle capable of autonomous operation, this has never been fully implemented. According to Sean O'Keefe the cost would not be a major show-stopper.

EELVs and CEVs

During the process of developing the new space policy, a preference was clearly evident in favor of retiring the shuttle fleet and using existing technology wherever possible to deploy the hardware needed to implement trips to the Moon and beyond. One way to do that is to try and use shuttle infrastructure wherever possible. The other alternative is to go with another existing capability albeit a rather new one that has not been fully matured: the Evolved Expendable Launch Vehicle (EELV). The Department of Defense, spurred to some extent by the repercussions of the Challenger accident and the cost of its aging Titan rocket fleet, paid Boeing and Lockheed Martin to develop a new, hopefully cheaper family of heavy launchers with the flexibility to adjust to varying missions, and to allow payloads to be developed such that they could fly on either company's vehicle.

Boeing created a wholly new vehicle in its Delta launcher family, which it had inherited when it absorbed McDonnell Douglas (which in turn had inherited the Delta from its original manufacturer Douglas Aircraft) to form the Delta IV EELV family. Similarly, Lockheed Martin added a new generation of launch vehicle to its Atlas line, which it had inherited with the acquisition of Martin Marietta (who had previously bought the Atlas' developer General Dynamics Space Systems division a few years earlier) to create the Atlas V family of EELVs. While the Atlas and Delta rockets inherited the name of legendary launch vehicles, they were wholly new rockets and were as similar to their predecessors as a new VW beetle is to the original design from the late 1940s.

The hope on the part of the government had been that a robust commercial launch market would help keep the price for government customers low. When much of the bottom fell out of that market, costs to the government went up and Boeing and Lockheed Martin were looking harder than ever for customers. While both rockets were designed to a rather stringent mission success criteria—ones similar in many ways to what a human-rated launcher would require—additional costs would still be needed to formally 'man-rate' the Atlas V and the Delta IV.

Rating these rockets for use with a human spacecraft, and modifying the launch facilities to accommodate human spacecraft would have to be paid by someone. Moreover, depending on the final design of the CEV, some alterations to the vehicles themselves might be required. Again the issue of who would pay entered the equation.

While the exact route NASA will take is still uncertain it would seem that the CEV will ride to orbit on one, perhaps both EELV families, should both still be in existence at the time. While the focus of this new space policy is toward an American agenda, it would be unwise to totally rule out potential international cooperation such that a CEV might find itself launched atop an Ariane 5. When asked directly by the authors if he could imagine a manned CEV riding an Ariane 5, Sean O'Keefe said 'why not?'

Cargo

The new space policy seeks to separate the launch of humans from cargo wherever possible. While the CEV will certainly have some minimal cargo carrying capability—suitable for carrying emergency spare parts, etc.—its lift capacity would be best likened to that of carrying the crew's luggage. Lofting of large payloads will be left to vehicles without humans aboard.

Currently the only way to get cargo to the ISS other than the shuttle fleet is the use of Russian Progress spacecraft; unmanned variants based on the Soyuz design originally developed for the Salyut space stations and Mir. In 2005 this capability will be augmented by the much larger ATV provided by ESA and launched atop an Ariane 5. At a date which is still somewhat uncertain Japan will also bring the HTV, its own robotic cargo ship, into service to the ISS.

Under the Space Launch Initiative (SLI) NASA had funded several companies to look at alternative ways to resupply the ISS. Known as 'Alt Access' this project was funded in fits and starts and was eventually terminated by NASA. Given the recommendations by the Aldridge commission, vis-à-vis more private sector involvement, and the CAIB's recommendation that crew and cargo be separated, it is likely that Alt Access will be resurrected in some form in the coming years.

Heavier cargo—cargo that exceeds the throw weight of an EELV—will have to ride on something else. Whether that heavy launch vehicle is shuttle-derived or a wholly new system is unknown. One thing is certain: the budget profile laid out for the new space policy does not have anything remotely close to the budget needed to develop a new launch vehicle.

As such, architectures will either need to be adapted whereby smaller pieces are launched on existing vehicles, or an additional pot of money will be required to finance the new launch system. While the new Moon and Mars architecture is just notional right now, one large payload is starting to move into a position of affecting thinking: Project Prometheus and the nuclear powered Jupiter Icy Moons Orbiter. Far too large for an EELV and possibly too large for a single shuttle launch, this payload is likely to be the tall pole that helps drive future heavy-lift decisions—until such time as other aspects of the new exploration initiative are translated into hardware designs.

One fact is unambiguous: America will not be building a successor to its venerable Saturn V moon rocket. Instead, the next Americans to walk on the Moon may find their ride made possible by a series of commercial satellite launchers bought in quantity.

Technology uncertainty

The future of the large U.S. expendable launch vehicle families will also be impacted by the Bush space policy. In its original policy construct, the EELV was to provide heavy-lift commercial and governmental launch services for twenty years. When the program down-selected to a pair of launcher families in December 1996, strong indications existed at the time that its eventual replacement in the far future would be some form of reusable launch vehicle (RLV), possibly one arising from the X-33/VentureStar technology demonstrations.

It is useful to reference a comment Lockheed Martin's Dave Urie made three years earlier, when the progenitor to Lockheed's eventual entrant in the X-33 contest, the AeroBallistic Rocket, was then being proposed in the halls of the Pentagon. Urie said that if that vehicle were brought to operational status, its success would depend on capturing all three elements of the launch market; civil, military, and commercial. In other words, a successful reusable launcher and a successful EELV could not exist in the same marketplace.

However, the technology path to such a future government RLV technology effort has been stillborn by Project Constellation. Both the Space Launch Initiative (SLI) and Next Generation Launch Technology (NGLT) programs were terminated. Two reasons were given for this action. First, NASA retrospectively claimed that the SLI was headed towards development of a lower cost launch system to service ISS payload needs. In contrast, SLI's Dennis Smith told a Space Transportation Association roundtable in March 2002 that configuration studies driving the SLI designs included payload requirements of the commercial satellite community in addition to NASA-unique needs.

Now, however, derivatives of either the CEV or some not-yet-developed commercial transfer vehicle would provide such a function, negating the need for the government to fund technology development of such a vehicle. Secondly, the cost environment facing NASA made continued spending on the launch technology packages represented by SLI and NGLT impossible.

Partners?

Another element in this process, a so-called partnership among NASA, the Defense

Department (DoD), and the U.S. Air Force to 'share' technology in launcher development, seems to have evaporated since the January 2004 Bush space policy announcement. One part of the joint efforts, the DoD National Aerospace Initiative (NAI), failed to receive funding from Congress last year. So it would appear that one should not look to the Pentagon for future extensive investments in affordable or reusable heavy-lift launch technologies.

The advanced engine programs once funded by SLI and briefly by NGLT, for booster engines and upper stages, are headed for a dead-end, following millions in investments by both NASA and its industry partners. NASA's primary contribution to the joint Integrated High Payoff Rocket Propulsion Technology program (IHPRPT) was contained in these two programs now headed for termination. Of the remaining IHPRPT partners, only the Air Force has consistently funded its share of the program, which was to lead to specific performance improvements in launch vehicle engines over the next decade. In the Constellation program, there appears to be no place for NASA's IHPRPT contribution.

As such, what then is the appropriate role of the federal government in funding advanced technology development for new rocket engines and upper stages in the Project Constellation era?

Without a follow-on launcher to the shuttles, and with the CEV not intended for large capacity cargo lift, another question may well be asked: Should the future of the EELV fleet be reconsidered? The primary purpose behind the EELV, driven by the DoD Space Launch Modernization Plan (the 'Moorman report') was to reduce launch costs by 25 to 50 percent over the existing vehicles of the time, mainly Titan 4.

While EELV flights have only just begun, it would appear that the two vehicle families have met that goal. What should the next step be in government funding for launch systems, post shuttle and into the middle years of EELV operations? With Congressional concerns expressed in June 2004 for reducing the current EELV sustainability to a single launch provider, it may be that O'Keefe and Steidle have but one domestic launch system to use to launch the Constellation astronauts. That is, unless a foreign system is employed. With the fate of the EELV family tied now not just to military lift but manned spaceflight, the issue of fleet survivability, as well as shuttle heavy lift, becomes that much more pressing upon the administration.

NASA's own track record in committing to development of any post-space shuttle launch system has been spending initial resources, followed by eventual abandonment for the 'next' great idea. None of these great ideas, however well they were advertised, ever yielded a launch solution.

A last question, then, for the administration might be: Why should industry believe that the launch systems developed for the Constellation program will be fully funded and supported across a shifting climate of political and space priorities? Why will this effort yield any different results than all those that have gone before?

If a new series of EELV designs are to be seriously pursued as candidates to bridge the gap between today's fully expendable launchers and a fully reusable commercially-capable RLV, who would pay the development costs? And could those costs be lowered to below the late 1980s' set of ALS-NLS-Spacelifter vehicle designs? That would require a long-term development plan for heavy-lifters, a roadmap for government technology investments, and a clear definition of the part that industry would need to play in this process. Such a roadmap would need to be a serious, sustained commitment for federal research and development for launch technology, more stable and dependable in scope than any of the pathways NASA or DoD has followed in the recent past.

A detailed space launch technology roadmap, however, is nowhere to be found in either NASA's post-shuttle plans or in DoD's post-EELV considerations. O'Keefe might well be moved to craft such a roadmap soon, while NASA is preparing for sending humans beyond Earth orbit. The value of using shuttle investments in boosters and tankage to sustain a heavy lifter for Project Constellation would appear to be no substitute for longer-range planning. Although efficient, it remains a stopgap solution that would temporarily preserve much of the current shuttle launch system workforce, provide for less development cost, and migrate the shuttle's operational competence during the Moon stage of Constellation. But what about the next stage?

A new launch solution for NASA and DoD?

During the next twenty years, when the evolutionary path to manned Mars missions approaches under the current timetable, the large lift requirements of both NASA and DoD could require a new launch solution beyond today's existing fleet. As the clarity of vision emerges as to the shape those mission models contain, so will the need for advanced space launch technologies made mature by government-industry partnership processes. A review of NASA's history in this regard suggests that may well be difficult to achieve, even if O'Keefe moves toward such a partnership as part of his transformation agenda.

In the end, it will not be today's space program leaders, such as Sean O'Keefe, Bill Readdy, or Admiral Steidle but future generations of Americans and others that will live with the ramifications of the framework of the retirement of the space shuttle system, or the replacement of the EELV, as well as the pacing of the transition to the CEV. Questions that arise from this framework should be asked before it takes final form and policy, for it will shape the evolution of human and heavy-lift space flight for decades to come.

Given the history of space flight, a cautionary tale is in order. Because what NASA does is ultimately more important than what it says.

ISS Costs: Finally Facing Reality

Several years before the new policy was set forth, the International Space Station had encountered major cost and schedule related problems. At the direction of the White House, starting with its FY2002 budget, NASA had been working toward a scaled back ISS configuration called 'U.S. Core Complete'. This stage in the deployment of ISS hardware would represent the capability required to support a crew of three—and be ready to accept partner hardware contributions from Europe and Japan—as well as some from Russia. This would occur with the delivery of Node 2. Any further development or 'excursions' as Sean O'Keefe was fond of calling them, beyond this capability up to the capability to support a 7 person crew (the 'baseline') would depend on how well NASA ran the ISS program.

According to a White House statement on the FY 2002 NASA budget. "Future funding decisions to develop and deploy any U.S. elements or enhancements beyond completion of the U.S. core will depend on the quality of cost estimates, resolution of technical issues, and the availability of funding through efficiencies within the 2002 Budget runout for Space Station or other Human Space Flight programs and institutional activities."

In addition, the White House ordered that previous ISS cost overruns would be "offset in part by redirecting funding from remaining U.S. elements (particularly high-risk elements including the Habitation Module, Crew Return Vehicle, and Propulsion Module). In addition, funding for U.S. research equipment and associated support will be aligned with the assembly build-up." Within months the Propulsion Module would soon be dropped after becoming prohibitively expensive. The next year the X-38, being developed as a crew return vehicle, would be discontinued as well—albeit in favor of what would later become the Orbital Space Plane concept.

The point at which ISS partner contributions, such as ESA's Columbus and Japan's Kibo modules, was referred to as 'Partner Complete', although the ISS would still only support a 3 person crew. 'Assembly Complete' referred to the point at which all of the baseline program's hardware elements have been launched and the ISS is ready to permanently support a 6 or 7 person crew. Much of the ISS configuration contained contributions by Russia (research science modules) which are not likely to be provided for Russian budgetary reasons.

It took NASA some time to get used to this direction—to focus only on U.S. 'Core Complete' and its 3 person crew. The White House had only committed NASA to building the ISS up to U.S. 'Core Complete'—yet many at NASA acted as if the work needed to get to a full 6 or 7 person crew was a given; and that it was simpler just to assume that was going to happen. This caused a lot of confusion since some people worked toward U.S. 'Core Complete' while others set their sights on a fully completed station.

IMCE: How to Fix the ISS Program

The most recent attempt at reigning in and focusing ISS research resulted from the recommendations made by the International Space Station (ISS) Management and Cost Evaluation (IMCE) Task Force. The IMCE was chartered to look in early 2001 to the causes of the $4 billion ISS program cost, revealed in Fall 2000. The IMCE's final report, issued on 1 November 2001, found a number of things lacking in the ISS program.

After nearly a decade of cost overruns and schedule delays, the IMCE was in a position to state the obvious such that something could be done to fix the ISS program. It found that "a cost of $8.3B (FY 02-06) is not credible for the core complete baseline without radical reform," noting that "the cost to achieve comparable expectations at assembly complete has grown from an estimate of $17.4B to over $30B. Much of this cost growth is a consequence of underestimating cost and a schedule erosion of 4+ years."

In making this judgment, the IMCE noted a significant schedule delay in setting a completion date for the ISS since the ISS program had been redesigned from the previous Space Station Freedom program. The FY1994 ISS program showed Assembly Complete as occurring in June 2002. The FY2002 program listed it as occurring in November 2006.

Focus

The IMCE would also make a series of recommendations that dealt with modifying the way the ISS program was managed, how costs were tracked, and how the program was aimed at actually accomplishing the tasks it had purportedly been designed to do in the first place.

Among the problems the IMCE found with the ISS was a lack of focus, why does the ISS exist, and what is it supposed to do?

Seeing the need to prod the development of a clear mission statement for the ISS, the

IMCE said at the very beginning of its report's preface, "A primary mission of the National Aeronautics and Space Administration (NASA) is 'To advance human exploration, use, and development of space.' We have developed this report upon this premise. The International Space Station (ISS) objectives require the establishment of a long-term human presence in space. A clear articulation of the mission of ISS within the broader context of the human exploration of space would greatly benefit the setting of research priorities for the station."

Among the IMCE's findings, it noted that "the Office of Biological and Physical Research (OBPR) is not well coordinated with the Office of Space Flight (OSF) or the program office for policy and strategic planning. The scientific community representation is not at an effective level in the program office structure." It recommended that NASA "develop a credible program road map starting with core complete and leading to an end state that achieves expanded research potential. Include gate decisions based on demonstrated ability to execute the program and Identify funding to maintain critical activities for potential enhancement options."

To address this issue, IMCE recommended that NASA "establish science priorities' so as to "give highest priority to research directed at solving problems associated with long-duration human space flight including engineering required to support humans in long-duration space flight." They went on to note that "the centrifuge is mandatory to accomplish top priority fundamental biology research. Availability as late as '08 is unacceptable." And that "research in the physical sciences of utmost importance can be accomplished on the ISS."

In addition, the IMCE recommended that NASA "establish a plan allocating available financial resources consistent with science priorities, including a prudent reserve," that NASA consider options to "augment 3-person crew" and look into "Extended duration shuttle" which might require "adding shuttle flights" or to "overlap Soyuz missions" so as to augment the number of crew on-orbit to do research. They also recommended that NASA "strengthen science management within the ISS Program."

In other words the IMCE sought to tighten up NASA's ISS research program, add financial and managerial resources, and provide some focus. But they were unwilling to reduce or eliminate non-human science that had already been selected for conduct aboard the ISS. In other words the IMCE wanted the original, broad, multi-disciplinary research program of the ISS to be implemented—as had been originally promised to users of the ISS.

This process of taking the existing program and tightening it up would be underway when the new policy sought to put the effort into overdrive.

ReMAP: Punting on Tough Questions

In March 2002, two years before the new space policy announcement, Sean O'Keefe responded to the research aspect of the IMCE report by creating the Research Maximization and Prioritization Task Force (ReMAP). ReMAP was chartered to "perform an independent review and assessment of research priorities for the entire scientific, technological and commercial portfolio of the agency's Office of Biological and Physical Research (OBPR) and to provide recommendations on how best the office can achieve its research goals."

ReMAP was composed mostly of space research veterans covering all aspects of research that was expected to be performed on the ISS. What followed was a large review process wherein all plans for the utilization of the ISS were reevaluated and compared against earlier studies which had sought to establish priorities. Despite the substantial amount of work that was put into the effort, ReMAP was not making any hard choices. Rather it was simply performing yet another review without any teeth.

On 10 June 2002 NASA formally announced that ReMAP had been granted an extension. The team had been scheduled to present their findings to the NASA Advisory Council (NAC) on 11 June 2002. The previous week ReMAP representatives had met with Sean O'Keefe. O'Keefe was not satisfied with what was presented and told ReMAP to go spend another month fixing the things that were not to his liking. O'Keefe's main concerns had to do with the way in which REMAP had prioritized various payloads and experiments—or rather, the lack of clear prioritization—with more than half of the experiments ranked as being 'top priority'. Without a clear stratification, it would be hard to prioritize the overall collection of work to be done and then map it against limited ISS budgetary resources.

The final report was eventually presented to the NAC two months later. Despite suggestions that ReMAP needed to sort research and payloads out more critically, such that those with the highest priority and those of lesser priority could be identified, that really did not happen. Since everything in ReMAP's final report was still more or less of high priority, ReMAP had failed in its main task. 32 Research priorities were listed as being first, second, third, or fourth priority. Only one discipline, 'bioinspired micro fluidics technology' was labeled 'consider termination'. In other words, ReMAP had simply rubber stamped virtually the entire status quo. Everything was of prime importance. Therefore nothing was of lesser importance. All REMAP had done was rearrange the deck chairs.

What is it that the space station is supposed to do?

Despite failing to 'reprioritize' ISS research such that tough decisions could be made,

ReMAP did acknowledge the lack of a central theme for ISS research saying in its report "the Task Force wrestled with the question of whether any further prioritization was possible; in particular, whether it is currently possible to place one of these broad goals at a higher priority than the other."

The report suggested that there were two broad paths ISS research could take—one which "addresses questions of intrinsic scientific merit (including those which might improve the human condition on earth) that cannot be accomplished in a terrestrial environment" and the other which: "obtains information necessary to enable human exploration of space beyond low-earth orbit, and develops effective countermeasures to mitigate the potentially damaging effects of long-term exposure to the space environment." While these two paths overlapped somewhat, the ISS was currently trying to do both—resulting in a very blurred picture when it came to defining why we had a space station and what goals it was constructed to meet.

The ReMAP report went on to make a cogent observation that it "recognizes the ISS is a truly remarkable facility that can be used to tackle either broad goal. However, the Task Force also believes that, while NASA's new Vision and Mission clearly articulate human exploration beyond LEO at some point in time, it is not yet known when such exploration might take place."

And of course, since ReMAP's charter was limited, as was the IMCE's, the ReMAP report concluded that "once the NASA time frame for human exploration of space is determined, research priorities that solve near-term problems can be distinguished from those that solve very long-term problems. Because the deciding factors for the next step are programmatic, rather than scientific in nature, the Task Force did not attempt further prioritization."

ReMAP did not question whether or not all of the existing lines of research should be done on the ISS. Rather, it went into an explanation why all of the existing and planned research was important and how it could be improved. No guiding principle was used whereby science was deemed to be supportive or non-supportive. To paraphrase a line from Lake Wobegon, "… all the space station research projects are above average."

Don't rock the boat

This wasn't really ReMAP's fault since the ISS itself did not have a central organizing principle that could be used to solicit and then select science research. Virtually all comers were welcome. If a peer review panel decided that your proposed research had intrinsic scientific merit and fit in somewhere with what had been called for in NASA Research Announcements (NRAs) you stood a good chance of being funded.

NASA was after 'world-class science'—something it wanted—but could never actually define. But NASA made certain that anything credible would somehow find room on board.

Strategy? What strategy?

Rarely in the 1980s or 1990s did NASA Headquarters life and microgravity science program managers think very strategically about what science they should be funding. They were not forced to ask themselves, "What do I have to accomplish by what date and therefore what science should I be soliciting?" It really wasn't their fault since there was no compelling reason to. NASA had no deadline whereby any strategic questions needed to be answered since there was no destination to go to; just the Space Shuttle and (later) the ISS going endlessly in circles. There was no pressing need to be economical or strategic; nor was there any long term funding or agency-wide direction focused on a specific goal. Given the lack of direction, managers focused on what they could reasonably accomplish with year-to-year funding and no big picture to guide them.

Moreover, NASA was averse to upsetting the cadre of researchers that had grown into a community within and outside of the agency. When budgets were cut at NASA, many program managers avoided canceling things and instead spread the pain around by cutting everyone's grants at roughly equal percentages. A more prudent thing to have done would have been to have all research ranked in priority and then start canceling grants so as to preserve the most important research, instead of diminishing everything. Again, NASA managers had no clear focus with which to make specific prioritized cancellations—so they didn't.

As such, NRAs had historically been written by NASA Headquarters staff to solicit proposals from mostly known entities doing existing research. If researchers at NASA field centers did not get what they saw as being their fair share they'd complain until they were satisfied. And if that route didn't work center directors would use discretionary funds to fill in. A classic case involved a lower body negative pressure device which was regularly rejected by NASA Headquarters peer review panels and then pushed back onto shuttle flight manifests by Johnson Space Center. And if university researchers did not feel that they got a fair share, Congressional staffers would be contacted and would then use earmarks to make certain money went where needed.

The net result was that little groundbreaking research was funded—since NASA wasn't really looking for it. If proposals passed peer review, they are funded, and that year's program was then constructed out of what passed peer review—instead of the other way around. Earmarks would then be thrown in on top and NASA would have to modify their program plan accordingly. NASA would also bend its science for political reasons as well. John Glenn's flight on STS-95 for 'scientific purposes' being the most egregious example wherein research already approved using the regular astronaut corps as subjects suddenly included a single geriatric research specimen and took on an aging justification.

As such, as the new space policy was delivered to NASA and called upon the agency to show some strategic thinking, it arrived at an agency desperately in need of a sharp, unambiguous focus to pull a meandering research program back from mediocrity.

Robbing science to pay for hardware

Research on the ISS was further hampered by the fact that NASA repeatedly took money that was supposed to have supported science research and payload hardware development to offset costs in building the ISS itself. According to the IMCE report: "When the Space Station Program was redesigned in 1993, the research support budget of $3.8B through assembly complete was programmed for research facilities and for recurring utilization. Between 1993 and 2001, the ISS Program experienced major delays, which resulted in slippage of the program schedule. As deviations in the program schedule occurred, the research support budget was realigned to keep synchronicity with the program. Consequently, the funding was taken out of the near-term years and was reinstated in the out years. During this process, the design, development, and fabrication of the research facilities were being delayed and experienced a cost inefficiency. This inefficiency, in combination with 4.5 years of inflation and $0.4B funding for Mir, has reduced the buying power of research funds by 40 percent. The total budget of $3.8B has not changed appreciably, but has been spread over a 13-year period for less capability. Discounting for the above factors (40 percent), the buying power of the current budget ($1.6B budget through FY02 – FY06) is approximately $2.3B."

ISS suffered from an attempt to satisfy everyone, and had to do so with a budget that witnessed a reduction in buying power by 40%. Regardless of what focus the ISS was given, it needed one. If not, then a substantial amount of money would need to be added just to make up for past raids on the science budget; to say nothing of new cost areas that had emerged in the meantime. Rep. Ralph Hall (D-TX) remarked in complete frustration in one hearing in the late 1990s that NASA thought it was perfectly OK to just come up to Capitol Hill, say that a certain budget would be spent on ISS research—all the while knowing in advance that the money would just be taken back to pay for other things. As such all of NASA's numbers with regard to monies set aside for ISS research were seen as being smoke and mirrors—just another bank account to raid.

Not only did the ISS need a clear research focus for planning sake, it needed one from a fiscal perspective as well.

Marching Army

One of the problems hampering the space station program, or any other large NASA program for that matter, is institutional inertia. The number of people and organizations that need something to do and the immense amount of effort required to implement substantial changes.

The IMCE report was rather blunt in addressing this issue, "The [space station] Program is being managed as an 'institution' rather than as a program with specific purposes, focused goals and objectives, and defined milestones. The institutional needs of the Centers are driving the Program, rather than the Program requirements being served by the Centers. The impact of institutional management is clearly indicated in the overall staffing levels of the program. The institution, not the program, controls the majority of these resources and timely destaffing is significantly hindered. At this phase of the ISS program, deleting more hardware saves very little money since the bulk of the expenditures are in the 'people' category."

You can't get a large, ponderous program like this to pivot in a new direction easily.

Taking the long view

Getting NASA to focus its science in a strategic fashion would not be easy. While the planetary and earth science portions of NASA were rather adept at this sort of task, the life and microgravity science communities had not been confronted with a pressing need to do so for decades.

Looking at an advisory report written in 1988 by the NASA Life Sciences Strategic Planning Study Committee 'Exploring the Living Universe: A Strategy for Space Life Sciences' you see a detailed suggestion for how to continue with the status quo for several more decades. Nowhere does the report seek to focus research towards providing a set of answers by a certain date in support of a specific goal—since the agency simply did not have any to offer.

With the Space Exploration Initiative (SEI) that grew out of President George H.W. Bush's speech a year later, a brief flurry of focused thinking of what research to pursue followed; but with mission dates so far off, little of the prioritization stuck. When the SEI evaporated in its own exorbitant cost estimates in the next few years, NASA's research agenda for the space station reverted back to an open-ended mode that it had followed previously.

Life-science research on the ISS (and previously on the shuttle) was clearly aimed at understanding how humans adapt to the space environment and looking for ways to ameliorate the deleterious aspects of that adaptation. However, no end point was given whereby the answers needed to be known, such that a mission to a given location could be accomplished. While some very insightful and valuable research has certainly been accomplished, it has taken much longer to accomplish than it might have had there been a sense of urgency in getting answers to specific questions. Meanwhile, microgravity and materials science on the shuttle and on the ISS was simply designed to see what could be learned—but there was no urgency to accomplish these endeavors by any specific date—if ever.

The Apollo program, although it asked a much simpler series of questions of its astronauts, methodically tested the limits of their endurance. They had to, since no one knew how they would react to certain situations and conditions and this knowledge was needed in order to design spacecraft and plan missions. As such, a structured set of projects, Mercury, Gemini, and then Apollo, incrementally tested the ability of humans to function safely and productively in space up to the mission durations required for a trip to the Moon. They did this with a rather clear set of objectives and a strong sense of the calendar.

Human research on the shuttle and ISS (and Mir) proceeded in an asymptotic mode, always approaching, but never actually arriving at an answer. This wasn't a case where the goal posts were moved to prevent someone arriving at an answer. Instead, there were no goal posts to begin with. In the case of Mir, the prime purpose was to learn how to work with the Russians in preparation for the ISS. Whatever science was done was of secondary importance. Indeed, given the $5 billion that the Shuttle-Mir program cost ($500 million for each of 7 flights plus a payment of $400 million or so to use Mir) a paltry amount of new science resulted. Indeed there is no reason why research being done on the ISS in the first decade of the 21st century could not have been done on Mir a decade earlier. Again, the reason was simple: NASA had no destination to officially get ready for.

As for the ISS program, NASA was still adrift when it came to what the station was supposed to be busy doing. The IMCE observed that the "Science program is not integrated in the ISS management" and that "Research implementation is proceeding assuming original program" such that there are "no defined science priorities," that Critical hardware [is] very late," the program has an "insufficient research reserve," and that a "3-person crew has [an] adverse impact on [the] science that can be performed."

In addition the IMCE observed that while NASA had been directed to focus on a 3 person capability and that it would later be determined whether it would be allowed to go to a full U.S. Core Complete configuration, that NASA was still working toward the 7 person configuration and that "the research support element of the ISS is still being implemented according to the original program" and that "the scientific community is confused and considers the reduction to a three-person crew, from the seven-person crew baseline, to have a significant adverse impact on science."

The other science

While NASA never wanted to formally admit it, the core purpose of its new space station (as had been the case with Skylab) was self-reinforcing—and always had been. The premise was that people would one day need to live in space for long periods as they went somewhere else. As such, NASA needed to be certain that they could do so safely. To do this preparatory work NASA needed a space station close to home where they could keep people aloft for long periods and work out how to keep them safe and healthy.

In other words the purpose of the space station was so that people could live in space. Many at NASA saw no need to justify this since they saw this as an inevitable activity for the human species. As such, the space station was designed, in one way or another, to flight certify humans for long duration spaceflight. Since NASA had no official destination, but still felt the need to prepare for one, it needed other things for a space station to do. Otherwise, probing members of Congress would ask why you need such a station if NASA has no official destination to prepare for. While not born a space enthusiast like his predecessors, even Sean O'Keefe bought into this, often saying of the ISS "if such a facility did not exist, we'd have to create it."

Therefore, the space station was sold as multi-themed research laboratory in orbit. Additional rationales used over the years are many, such as inspiring youth, etc..

Rep. Tim Roemer (D-IN) and Sen. Dale Bumpers (D-AR) long-time space station foes, would often regale their colleagues with a long list of everything NASA ever suggested as being the purpose for a space station and then wave it in NASA's face each year as the space station seemed to have lost one bit of capability or decided not to pursue one thing or another. Each year attempts to kill the station would fail, but sometimes they'd get really close. NASA had sold the space station as all things to all people. Any decrease in capability was therefore seen as somehow compromising its intrinsic value. This notion caught on because NASA never managed to give the space station a simple task.

To be certain, much of what was proposed to be done on the ISS truly had clear, intrinsic merit. The space environment is an incredibly novel and interesting place to do things. Microgravity is a condition that can only be replicated for moments in close proximity to Earth. By removing it for long periods of time, you have an unprecedented opportunity to understand a number of underlying principles of many physical phenomena without the masking effects of gravity. Some forces which have little effect on the physical world on Earth are free to prevail in space—often with surprising results.

As to what the benefits are of such research—as presented by NASA—they ranged from a simple understanding of how the universe works to over-hyped promises of the production of wonder materials and miracle drugs. Sadly, it was often the promise of miracle cures in space or some dramatic advance in metallurgy—promises often repeated without question by Congressional supporters during annual floor debates—which would wear thin years later when such advances never appeared.

Given that NASA was already going to be studying humans in space, there would be people up there. Why not use this on-orbit resource to do other research as well? It was a no-brainer to add a program of microgravity and materials science to the space station's research portfolio. But you did not need any of this to understand how to keep people in space for long periods of time.

In addition to microgravity and material science, there was a lot of non-human biology you could do, also using the novel microgravity environment. Gravity is the one environmental factor whose strength and direction has remained unaltered throughout life's tenure on Earth. As mentioned before you cannot replicate its absence for more than a short time near Earth's surface. Indeed, gravity is the one environmental force for which life, in its entire existence on Earth, has never had to evolve adaptations to its absence. And no, floating in water doesn't do that. Nor does NASA have any zero gravity chambers. Putting living things in space is a wholly novel experience and doing so can be very illuminating as to how living systems work.

As with microgravity and materials science, there are some fundamental issues—one of them being life's response to gravity—that can be answered. But many of these research questions do not have a direct link to human survival and productivity. Other types of non-human research have some application—especially those where sacrificing animal specimens in order to study tissues is required.

To spin or not to spin

Some of the most complex research facilities planned for the ISS—furnaces and the centrifuge facility—are required to support mostly non-human research. One can argue (and NASA has argued) that using a centrifuge could be part of a program to understand whether artificial gravity is required for long voyages such as those to Mars, and that studying its use on animals would be a good way to gauge its utility. Yet evidence from long-term stays on Russia's Mir and the ISS suggest that 'centrifugation', a very expensive solution, is probably not required so long as the crew conducts various exercise countermeasures.

Astronaut Michael Foale returned to Earth in April 2004 after six months on the ISS, a mission duration equal to a trip to Mars. A few weeks after his return, Foale said that he felt that he would have been able to get out of his Soyuz and conduct surface science in Kazakhstan hours after landing if he had been called upon to do so. Such would be the case when astronauts landed on Mars. Moreover he said that after his six trips in space that he felt that artificial gravity would most likely not be required for trips to Mars.

The IMCE said that it felt that the centrifuge facility was important, saying, "The IMCE recommended that NASA 'Provide the Centrifuge Accommodation Module (CAM) and centrifuge as mandatory to accomplish top priority biological research. Availability as late as FY08 is unacceptable...'"

However, the IMCE made this recommendation in the context of looking at how to support the original, broad research agenda that the ISS had adopted. Looking at a space station which now has Human-centered research as its core focus puts the centrifuge in a new light. Suppose the U.S. were to decide that existing data (such as Foale's experiences) suggested that a microgravity trip to and from Mars was acceptable in terms of the risk to human health and performance. In addition, assume also that NASA then focused its space station research on how to take existing human zero-gravity exposures and better study crews for even longer periods. Then the rationale for the Centrifuge Facility is reduced substantially.

Much of the operation of the Centrifuge Facility is intended to support basic biology studies not related to human health. In such experiments, the Centrifuge is to be used as an on-orbit experimental control to be certain that observed effects are due to microgravity or spaceflight factors. Dropping basic biology research not directly linked to human health from the ISS research program would cause the need for a Centrifuge Facility to all but evaporate. This is likely to be a heated discussion given the long history of pushing for the Centrifuge Facility on the ISS and the use of similar, but smaller systems in other spacecraft. Nonetheless this is going to be but one of the very difficult choices NASA is going to have to make.

NASA transferred the development of the Centrifuge Facility to Japan in the mid-1990s as part of a barter arrangement. In exchange for building this facility (which would still be considered U.S. hardware once delivered) Japan was able to offset costs associated with the launch and operation of its Kibo laboratory module. Not to fly the Centrifuge Facility now would upset that barter arrangement. Given that many within Japan view the long-term value of its space station participation with increasing skepticism, such a decision would only serve to deepen their concerns.

In addition to tough choices in life sciences, if microgravity and materials science are no longer to be accomplished, also because they do not contribute to the human science agenda, then the need for the furnaces and other associated hardware similarly disappears. As is the case with non-human basic biology research, a large majority of those involved are resident at universities spread out across the U.S.. Each grant cut by NASA results in complaints to at least one Congressman and two Senators and they defend these researchers fiercely. Hundreds of grants are at risk. Do the math.

Hard choices for science

At a June 2004 meeting of the NASA Advisory Council, NASA's Associate Administrator for the Office of Biological and Physical Research, Mary Kicza presented an update on the changes within her organization. Charged with coordinating life science and material science research for the agency, OBPR faced a major sea-change with the announcement of the new space policy. Gone would be the days when any science of intrinsic value would be welcomed aboard the ISS. Instead, all research would now be focused on human exploration issues - health, life support, productivity, etc..

Kicza noted that her office had sent out several communications with its research community, a 'Dear colleague' letter, a week and a half after the president's announcement, followed by two subsequent emails. This was followed by 6 workshops, 2 internal NASA reviews and 4 external reviews with external participants including the National Research Council's Space Studies Board, Aeronautics and Space Engineering Board, and the Institute of Medicine.

Once decisions on program changes are made letters will be issued to individual researchers in July 2004 and a broad 'Dear colleague' letter will be sent to the community as a whole, outlining the changes.

One of the core planning activities to be modified for use in implementing the new space policy is OBPR's Bioastronautics Critical Path Roadmap. In light of the new policy this document, first assembled in 1997, underwent considerable revision and was opened up for community review and comment. Interest in the document was significant and the deadline for comments was extended to accommodate.

According to a cover memo sent out with a request for comments, "One of the major changes was to align the BCPR with the Agency's new vision for the exploration era. We expanded the context for risk identification and assessment to include three Design Reference Missions (DRMs) (a one-year stay on ISS, a one-month stay on the lunar surface, and a 30-month Mars mission). Other changes included: streamlining the content, eliminating redundancies, and including a greater representation of human support technologies and autonomous medical capabilities. While additional risks in the latter two disciplines were identified, many of the original human adaptation and countermeasure risks were consolidated. In addition, many more exploration enabling research and technology questions were delineated, and notional schedules were developed to illuminate important research activities and milestones. The current BCPR reflects a more integrated framework for risk reduction and management."

While this effort led to a further refinement of human research needed, spurred on by the first overt commitment to specific destinations in more than a decade and a half, other aspects of OBPR's programs would suffer in contrast.

In the NAC meeting Kicza said that OBPR has "identified projects not aligned with [the] Exploration vision" and that this included "9 or more major hardware builds, 2 Express racks, 25 flight investigations, 50 ground based investigations" and that "[a] funding transition strategy [is] planned for investigations." Before any dramatic changes were made Kicza sated the obvious—that there would have to be 'congressional input'.

Among the alternatives Kicza was considering: outright cancellation, putting the project on hold, converting what had been a program to fly experiments in space to one that did research on the ground, or find alternate ways to fly things in space, perhaps aboard a new series of unmanned satellite. Curiously, this was not NASA's first flirtation with the idea of developing small dedicated life-science satellites. In the late 1980's a concept called 'LifeSat' was proposed that would allow a series of experiments to be done before the space station was operational and afterward so as to complement the research capabilities aboard the station. The project was killed for political reasons, the most notable being that this was seen as undermining the rationale for the space station in the first place i.e. if you could do science on low cost free-flyers why did you need an expensive human-crewed space station. Once again the political winds had shifted and free-flyers were now seen as being valuable.

Kicza was pushed repeatedly during the NAC meeting to be specific as to what was being canceled. NAC member John Glenn, a frequent critic of the new space policy brought up the ReMAP comment that science should be parsed into two general categories—human and fundamental. It would seem, according to Glenn that the fundamental science was taking the brunt of the proposed cuts. He also noted that cutting this fundamental science would undermine, in some people's eyes, the reason why the space station was built in the first place. Kicza made repeated reference to the fact that these decisions were being made in response to a policy handed to the agency by the White House to which Glenn quipped, before the NAC, "Did not most of the directive come from over here (NASA)?"

Kicza responded with greater specificity. While not taking Glenn's bait, she did say that low-temperature condensed matter physics research was not going to fly. Other research such as fluid phase transformation and integrated cell/genome studies might fly in a modified fashion. A week later, after a request was placed with NASA Public Affairs, NASA refused to provide any further detail on the projects it had identified saying, "Affected projects will be determined with more certainty and announced publicly after completion of the fiscal year 2006 budget process and proper notification of Congress, which will take place over the summer."

As such, NASA was going to do a modern day equivalent of what King Solomon threatened to do with a child when two women claimed it as their own. It would sacrifice its own. But it would not say how this would be done until it was good and ready. This would not go down well with the science community who had stuck with NASA through thick and thin supporting the need for a facility to do a broad range of research.

Moreover, many members of Congress who had made long impassioned speeches lauding the merits of basic biology research, protein crystallization studies, and material science research on the ISS. They had often done so in the face of blistering criticism from those who saw money spent on anything BUT space. These members of Congress would now find all of their past rhetoric to have been for naught and somewhat embarrassing. Soon after, they would get angry—not a good thing to have when you need all the friends you can find in Congress.

With the announcement of NASA's new organizational structure, NASA's Office of Biological and Physical Research was absorbed into the Office of Exploration Systems. To those who felt that such a close integration between the human aspect of space exploration and the systems needed to make it happen, this decision was welcomed. To those who felt that this subjugated basic science to the pressures associated with mission operations and architecture, the decision was greeted with skepticism. And of course, to those whose projects would soon die, the decision was greeted with anger and frustration. Frustration that goes with pouring yourself into a project for a good portion of your professional career only to see the rug pulled out from under you.

The Partners Aren't Happy

One of the consequences of NASA's FY 2002 budget guidance was to pull the rug out from under the partner nations that had already joined with the US in the ISS program. This wasn't the first time. In 1992-93 the Space Station Freedom program (already an 'international' space station program) was redesigned into what would be called the 'International Space Station' with the addition shortly thereafter of Russia. At that time the capabilities of the station were decreased and the Russian participation more or less forced upon Japan, Russia, and Canada.

Now, a decade later, with the adjustment of the program and focus on a three person capability, the partners once again felt betrayed. The IMCE observed, "The International Partners believe they have been told the original program is the baseline and that they had 'Treaty level' commitments," to back up this position. Once the IMCE report was issued and agreed with the White House's plans, formal diplomatic complaints from Europe and Canada were made public. Japan took less overt methods to make their concerns known.

The ISS program was operated, as was the SSF program, on a basis whereby participating nations donated hardware to the project, a fixed portion of overall ISS resources (utilization) was generated based on the size of their overall hardware contribution. Using this percentage, nations would also pay a portion of the overall operating costs of the ISS. With the exception of operating costs, little if any exchange of funds between nations occurred. Subsequent alterations to the content of the program were done by 'bartering' of hardware, crew times, and other resources between nations, to meet evolving needs. U.S. fees were paid to Russia for some things. This was the only notable exception—and was brought to a halt by the Iran Nonproliferation Act.

The shared utilization percentage for ESA, Japan, Canada, and Italy is spread across the entire U.S. on-orbit complement—hardware as well as crew time. Use of Russian facilities follows a somewhat different process. Conversely, the U.S. had a set utilization percentage within the facilities provided by ESA, Japan, Canada, and Italy. These agreements all assumed a baseline of a 6 and then later 7 person crew. The 7 person crew capability was offered by NASA when it was planned that the X-38 was to be used to replace two three-person Soyuz crew return vehicles. The final number tended to revert back to six once the X-38 was canceled.

With a three person crew on the ISS, and then taking deductions for overall ISS maintenance, much less time for science would be available than would be the case for a six person crew. One can argue about how to calculate crew productivity, how to enhance efficiency, what a work week is, etc. but simple math tells you that having three people is less than having six people. This benefited no one who had assumed a 6 person crew when they did their science planning. The IMCE stated, "The U.S. Core Complete configuration (three-person crew) as an end-state will not achieve the unique research potential of the ISS."

ISS utilization at the point at which the ISS was complete – i.e. the aim point for full-fledged science, was 6 people. This is what everyone signed up for. To now suggest that either 3 people is the end point, or that the 6 person capability is going to be substantially delayed was seen by the partners as going back on an agreement—one they held in very high significance.

Of course, other than crew time, with the wholesale refocusing of all that the U.S. was contributing to the ISS was going to complicate things further. With the announcement of the new space policy in January 2004, the U.S. now wanted to move the U.S. utilization of the ISS from a broad range of capabilities and disciplines, to one: human life science. This increased the complexity of the partner's unhappiness. The deletion of materials and microgravity science work, and thus the glove boxes, furnaces, and other support hardware the U.S was going to provide, would force Japan and Europe to rely on only what they had decided to provide or from new research alliances with each other. They would also have to face the prospect of U.S. researchers, with interests in areas the U.S. was dropping, knocking on their doors looking to collaborate.

As NASA moved toward the anticipated release of the Aldridge Report, all space station researchers were nervous. And waiting.

*   *   *   *   *

11

TO WORLDS BEYOND

Visionaries

On the upper floors of a generic office building a hand-picked group of men and women—some youthful, others with their faces lined from years of experience—all hunch over computer screens on desks stacked with reports. This small group moves about their offices, cubicles, and conference rooms with dispatch and a quiet sense of urgency. They also have a spring in their step. They have a vision to accomplish; even if the rest of NASA has not quite gotten the message yet.

The office building is NASA's Washington, DC Headquarters—a location not usually known as a center of innovation.

Since President Bush came to this building on the cold winter afternoon of January 14th, much attention has focused on the controversial political issues facing the new vision for space exploration given to the civil space agency. But while the politicos and pundits battered the Bush initiative on Capitol Hill and in the media, NASA began to quietly chart the course for the hardware needed to implement the vision— a course that hopes to speed up the normal slow-paced process of defining spaceship designs and issuing contracts to create them.

On their plate: Project Constellation—which will develop the Crew Exploration Vehicle (CEV) —the first wholly new human space transport created in a generation; Project Prometheus which seeks to renew America's embrace with nuclear propulsion to reach destinations faster and with much more capability upon arrival, and a series of missions which will send robots ahead of humans as America looks back at the Moon and then on to Mars.

Leading the charge

Sean O'Keefe brought retired Admiral Craig E. Steidle to NASA to be the new Associate Administrator for the newly created Office of Exploration Systems (OExS)—an appointment made at the same time as the president's vision was announced.

Steidle wasted little time jumping into the task.

Steidle arrived with a background that included managing the Joint Strike Fighter program—in a manner that won his team the 1997 David Packard Excellence in Acquisition Award. Their efforts were focused upon the creation of a common family of aircraft; each with a different role. According to a press release issued at the time, this team's efforts resulted in "development savings estimated at nearly $18 billion (in FY 95 dollars) and life cycle cost savings projected at 33-55 percent compared to historical aircraft programs." In addition to cutting development costs, the JSF program under Steidle also sought to aggressively manage technical risk prior to entering into the engineering and manufacturing development phase of the program. Lastly, the JSF effort enhanced competition by "sharing an unprecedented amount of key process information with competing contractors."

Steidle also managed the Navy's F-18 program, and directed development of the F-18E/F 'Super Hornet'. He was a decorated Navy pilot, who had flown over 50 different types of aircraft, including the F-18, A-6, F-4, A-3, and H-2. In so doing he accumulated over 3,600 flying hours including nighttime carrier-based missions over North Vietnam.

Given NASA's chronic problems with delivering programs that met initial requirements on time, and managed to arrive within reasonable distance of cost predictions, someone like Steidle was exactly what NASA needed if it was going to make this latest challenge happen.

Steidle was given a lot of managerial latitude by O'Keefe. As such, he was able to tap resources within and outside of the agency, as needed, to set up his organization. The net result after six months was a mix of some familiar faces and some new ones. Unlike some of the more established organizations, this team had a sparkle in their eyes as they presented their plans and chronicled their progress. Indeed when their logo and motto rolled out a few eyebrows were raised. 'Audentes fortuna juvat' on the left and its English translation 'Fortune Favors the Bold' on the right. A little of the 'old NASA' seemed to be stirring.

Business unusual

Standard NASA practice on such a big, new endeavor would be to have a big splashy press event—one complete with lofty goals, some pretty pictures, and some managerial announcements, followed by some obligatory Congressional introductions. Everyone would then fade back into the woodwork for a year or so as they worked to figure out just how the heck they were going to transform pretty pictures into hardware.

Steidle wouldn't have that—he couldn't. NASA was in the hot seat almost as soon as

Congress had a chance to pontificate. They wanted to know what the CEV would look like, how much it would cost to do all of the things it was supposed to do, when it would go to Mars, etc.. They wanted answers yesterday. Given NASA's track record, it was hard to find fault with their concerns. But it was also a bit naive to expect this much detail from a project that had only been a decision package on the president's desk a few weeks earlier. The agency needed a go-ahead from the president before they began a crash effort to implement the plan.

Besides, contrary to what many would think as they looked back fondly to Apollo, it took NASA several years to get its Apollo mission architecture into place. Indeed it was another 14 months after John F. Kennedy's May 1961 speech at the State of the Union before the Lunar Orbit Rendezvous mission approach was formally adopted by NASA—and Apollo architecture trade studies had already been under way for more than a year prior to Kennedy's speech.

In a reflective echo from the past, one of Wernher Von Braun's initial approaches for making a lunar landing—one that was rejected at that time—Earth Orbit Rendezvous, was being looked at anew. Why? Because it would not require building a huge new launch vehicle to send the CEVs out from Earth orbit. Rather it could instead use versions of today's Delta IV, Ariane 5, Atlas V or an unmanned cargo version of the space shuttle, to assemble the Moon bound Constellation fleet in Earth orbit. What is old is sometimes made new again.

Historical corrections aside, that was then and this is now. As such Steidle had to hit the ground running—and not falter. All the while he'd be watched in real time with considerable transparency into his evaluation and decision processes.

Where's the eye candy?

During the last days of the Space Launch Initiative (SLI), timelines were presented by the SLI program office managed out of Marshall Space Flight Center which showed the development of the Orbital Space Plane and other vehicles intended to augment and/or replace the Space Shuttle fleet. At the same time, the NASA headquarters Office of Space Flight, the obvious customer for such systems, showed charts with the shuttle fleet flying for decades out to 2020; but made no mention of OSP or other SLI systems. When this disconnect was raised with either organization they'd either say that the missing detail was not their responsibility or grumble and say, "No kidding."

In addition, many dozens of pretty computer graphics showing a swarm of different launch and upper-stage concepts would exude from SLI and all of the contractors involved. Looking at all of this eye candy, the casual observer might think that NASA really had a lot of things going when in fact none of the systems depicted in these images were real. They were just ideas. X-33 and X-34 had started that way too; pretty pictures and empty NASA promises.

Some progress had been made since then. With the new space vision you now had the Crew Exploration Vehicle and other possible launch systems designed to augment and eventually replace the space shuttle and you had the International Space Station as a test-bed for exploration research. Yet, with the exception of one generic 'big picture' chart which showed all of the future bits and pieces—a chart unaltered since its presentation with the president's announcement—the Office of Space Flight and the Office of Exploration Systems still seemed to avoid making overt mention of existing and new systems in the same breath—but they wouldn't make funny faces— and admit that you had a valid point if you brought this to their attention.

In addition, the number of pretty spaceship pictures was smaller this time. Within a short time of Bush's announcement Boeing and Lockheed Martin made public some online galleries of concepts which looked very similar to each other; most of which were derived, some with little alteration, from artwork they had developed for their Orbital Space Plane proposals. Other than that, NASA did not go off and commission nearly as much conceptual artwork as it might have done in the past. Yet people still wanted to see what the moonships looked like, but NASA was somewhat shy about showing any examples.

Steidel was asked at the June 2004 industry day when the 'big picture' would start to percolate further into the details of what OExS presented. Moreover, when would people get to see what concepts NASA was considering? Steidle replied as if he'd been asked this eight times already that day, "It is critical that we start with a strategy with clear ideas without specific outcomes pre-judged." Steidle wanted his technology program to resolve the challenges and to solve them he had left the solutions as an open book. He wasn't going to force a solution down anyone's throat. Rather, there was going to be an iterative process where agreement would be bought by the yard on each step down the path toward consensus such that the final solution would be one that had been derived—not imposed. "We do not want to define the solutions ourselves." Steidle would add. "We want to use a larger group." According to Captain Michael Hecker, Steidle's Deputy for Development Programs, "We are going in with no preconceived solutions."

As part of this process Steidle said that his office was purposefully careful not to put a bunch of pictures out. There was a certain practicality to this. As far as your typical NASA (or Defense) project goes, the more you put out publicly in terms of pictures, definitions, mission profiles, timelines, etc. so as to sell the program, the more risk you'd run down the road of having to modify those early impressions when your program falls behind. Expectations become inflated. The situation gets stickier if you have a bunch of politicians who have stood up for you based on the early hype who now have to eat some crow because you have failed to deliver.

Of course, if you are frugal about pumping this stuff out at the onset it becomes more difficult to sell the program and to gain supporters. Steidle seems willing to take that risk as a trade-off for a program which, once implemented, is far less likely to fail to meet public expectations, since the public face given to the project has undergone much more thought and scrubbing prior to launching and PR promotion.

The progress is in the eye charts

Pretty pictures notwithstanding, Steidle was not at all lacking in details to flash on the screen. Within a short period of time, as he added staff and formulated his initial plans, detail began to be added to the overall scheme he would paint. By the time the Aldridge Commission would issue its report, Steidle had arguably made a year's progress (perhaps two) in six months time—if you were to look at how NASA usually did such things. Moreover, what he had accomplished was crisp, taut, and to the point—again something NASA traditionally had problems achieving.

In a presentation to the NASA Advisory Committee in June 2004, Steidle presented a status report on where his organization found itself after six months of operation. Steidle had made repeated comments prior to June that he intended to learn from past findings and lessons learned in so far as how NASA—and other organizations—mounted missions. Referring to advice offered by the Packard Commission with regard to reducing cost 'Overruns on Defense Acquisition Contracts' he cited the following as what he and his organization were building upon, "Get operators and technologists together to enable the leveraging of cost-performance trades; apply technology to lower costs of [a] system, not just to increase its performance; mature technology prior to entering engineering and systems development; and [form] partnerships with industry to identify innovative solutions." In addition Steidle cited the Young (IMCE) report, "Requirements definition and control are dominant drivers of cost, schedule, and risk in space systems development programs."

A week later OExS sponsored an "Industry Day" which was held in a large auditorium at the Commerce Department in downtown Washington. This venue was chosen because the 400+ seat Webb auditorium at NASA Headquarters was filled to overflowing for a previous event. Everyone had a seat this time but there were few empty ones left. The attendees were a who's who cross section of the aerospace industry. The event was timed to precede the release of two Broad Agency Announcements (BAAs) which would be the first contracts awarded by Steidle's team.

What's different this time?

At one point a panel of OExS staff was assembled for questions from the audience. The panelists were asked what the key difference was between this new effort, the president's space 'vision' and NASA's previous large effort, the Space Launch Initiative—with the suggestion that that this new effort seem focused while SLI was a grab bag of different projects. The panelists all shifted in their chairs, hesitant to answer. Then they bit the bullet and replied.

"This time we have a vision," one panelist said. Astronaut Brent Jett said, "The key piece is that we have a program laid out over the long term with near-term goals that are well defined." In addition systems being developed for initial use on the Moon will have 'extensibility' i.e. the ability to have inherent capability to be used to create hardware and systems needed to go elsewhere—including Mars.

Garry Lyles who had managed the Next Generation Launch Technology portion of SLI said that spiral development was an important change as well. "You don't have to tackle the job for the next 30 years—you can look at the first spiral—that helps you 'crispen things'. We did not have that with SLI."

Another difference cited was the fact that the requirements in this program would not change. One thing Craig Steidle is eager to mention is a pledge regarding requirements: "We're going to lock down requirements and I am not going to change them on you."

A new mantra

Steidle said from the onset that he was going to learn from NASA's successes and mistakes, and implement a plan in a way which would soon be coined in a new agency mantra 'affordable – sustainable – credible'. Mantras can lull people into a false sense of security and eventually get NASA into trouble—and you have only to look at the previous agency motto to see just how much hubris was embodied in this new one.

During the June industry Day long time NASA advanced policy guru John Mankins—a forward-looking survivor who had managed to keep the faith during the Goldin years—summed it up perfectly, "We don't want to get into the situation we were in a decade ago with the agency's 'better-faster-cheaper' mantra—one wherein people were told you can have any two of the three—but not all three."

No one could ever explain what faster-better-cheaper meant. Given the chronic budget cutting hailed and encouraged by Dan Goldin for nearly a decade everyone at NASA soon learned what it really meant by default: 'cheaper'.

Perhaps the most telling example of this interpretation can be found in the September 1999 issue of NASA JPL's 'Universe' newspaper. Sam Thurman, flight operations manager for the Mars Climate Orbiter (MCO) at JPL was clearly under the mantra when he said, "If you look at the number of individual engineering jobs we have to do in this six-month period and how the Orbiter and the Lander interact to accomplish their respective missions—all with a team of 80 people to do it—to me that's where we're breaking new ground. It's going further and faster with fewer people and with a smaller budget. If we're successful, I think we'll raise the bar on the whole faster, better, cheaper mantra to a new level—to a level that's not been attained by anyone else."

"Faster" and "cheaper" were in evidence. "Better" was nowhere to be found. Two spacecraft slammed into Mars. In once case (MCO) when the process of converting metric into English units was not done; because there was no one to make sure that it was done. Mars Polar Lander crashed because there was not enough money or time to do a full systems check and the force of its landing legs deploying told the lander that it had already landed—before it had indeed landed—and its engines were shut off. Stupid, avoidable mistakes.

As the agency emerged from these back-to-back harsh lessons-learned sessions, it would be slapped down again after the Columbia accident. Now Steidle's organization was going to attempt a long-term integrated human and robotic exploration effort. It had two hurdles to surmount in order for this to happen: the engineering inherent in building the spacecraft and the equally formidable task of reversing the tarnished reputation of the agency at the same time. In so doing everything would need to be 'affordable – sustainable – credible'—a mantra of pragmatism.

Requirements first and foremost

Craig Steidle placed a strong, 'no exceptions allowed' emphasis on getting the requirements locked down early—and then protected those requirements from change thereafter. Requirements changes, 'requirements creep' as it is often called, is the bane of NASA's existence. Back in the 1990s when the X-33 and X-34 had problems meeting their required goals, NASA reduced the goals until such time as the hardware to be produced was not worth continued investment. A billion dollars down the drain. Had the requirements been scrubbed before the program began—for their ability to be met—and a sanity check made by all involved, including the contractors, a more reasonable program might have been planned and implemented. Back then NASA was in a time honored mode of sell the program first—and then figure out how to do it later—and if need be ask for more money, reduce or 'descope' the requirements, or stretch out the schedule (and drive up the cost).

The International Space Station Program, and its predecessor the Space Station Freedom Program, routinely modified requirements well after the design process began—either during crash redesign efforts (which seemed to pop up like daffodils every spring) —or during non-rush periods. All of this happened as design reviews were underway in parallel. The net result was often a time lag between requirement change and the ability of the program to respond by altering hardware designs. After going on for a number of years the program was in a constant state of catching up, with liens and known disconnects, to the program's design. Adding in the Russian components and their 'heritage hardware' designed to wholly different safety and reliability requirements made things worse. This eventually resulted in hardware in space which did not meet acoustic, debris impact, or other operational requirements; things which had to be fixed after the fact, in space, at additional cost.

Steidle had seen similar things for himself during his Navy days. He also had no shortage of NASA veterans to tell him horror stories. In addition he had seen how to avoid this situation during Joint Strike Fighter development. This program he would now run at NASA was going to be different than anything the agency had done in a long time. At the same time it would be familiar to those who had seen NASA pull off its successes as well.

Starting from Zero

The first task was to develop the requirements and once everyone had done their task to cast them in stone as a solid basis upon which to build a program.

Deputy Administrator Fred Gregory gave a presentation to the NASA Advisory Council on the requirements definition process for the new space policy in June 2004. This process was, to begin with, a simple set of things that the agency was going to do; things which were directly descended from the president's January 2004 speech, and simple enough that they fit on a single sheet of paper. A similar approach had been mandated by Sean O'Keefe in 2002 when the high level requirements for the Orbital Space Plane were developed. NASA would take the lessons learned from that process and apply them here.

To develop these 'Level Zero' requirements, NASA assembled a series of teams to parse through the process. The 'Core Team' operated from 23-27 February 2004 and used a 'from Mars looking backwards' exercise to take an initial stab at 'plausible scenarios and products'.

Shortly thereafter, for two weeks in March 2004, Gregory said that a 'Blue Team' worked to develop "a strategic architecture framework, Level Zero requirements, and conducted a 'from today looking forward' exercise to determine needed capabilities in the required timeframes." The output of the 'Core Team' and the 'Blue Team' was then run through an 'Integration Panel' and the result of the integration fed in to a 'Red Team'.

The Red Team, composed of 'senior agency officials' spent several days in Mid-March and "evaluated architectures, needed capabilities, Level Zero requirements, and near-term forward work and assessed implications for agency skills and capabilities, and provided recommendations to the JSAC and the NASA Executive Council." The JSAC (Joint Strategic Assessment Committee) then reviewed the outcome of the Red Team on the following two days so as to develop Level Zero requirement recommendations for the NASA Executive Council to review and approve.

These Level Zero requirements were also coordinated with the White House and the Office of Science and Technology Policy (OSTP). Once that review was completed, the Executive Council formally established the Level Zero requirements. These requirements were the implementation, in requirement format, of the president's policy as originally outlined on 14 January 2004.

This is the simple codification of the president's direction they arrived at and baselined on 21 May 2004. Note that if Craig Steidle has his way (and subsequent presidential administrations concur) these words will be the basis for American space efforts for the next several decades:

Executive Council Approved Level 0 Exploration Requirements

Mission Statement

NASA shall advance U.S. Scientific, technological, security, and economic interests through a robust human and robotic space exploration program.

Level 0 Exploration Requirements

NASA shall implement a safe, sustained, and affordable robotic and human program to explore and extend human presence across the solar system and beyond.

(1.1) NASA shall develop the innovative technologies, knowledge, capabilities, and infrastructures to support human and robotic exploration.

(1.2) NASA shall conduct a series of robotic missions to the Moon to prepare for and support future human exploration activities.

(1.3) NASA shall conduct human lunar expeditions to further science, and to develop and test

new exploration approaches, technologies, and systems, including the use of lunar and other space resources to support sustained human space exploration to Mars and other destinations.

1.4 NASA shall conduct robotic exploration of Mars to search for evidence of life, to understand the history of the solar system, and to prepare for future human exploration.

(1.5) NASA shall conduct human expeditions to Mars to extend the search for life and to expand the frontiers of human exploration after successfully demonstrating human exploration missions to the moon.

(1.6) NASA shall conduct robotic exploration across the solar system for scientific purposes and to support human exploration.

(1.7) NASA shall conduct advanced telescope searches for Earth-like planets and habitable environments around other stars.

NASA shall acquire an exploration transportation system to support delivery of crew and cargo from the surface of the Earth to exploration destinations and to return the crew safely to Earth.

NASA shall complete assembly of the International Space Station, including the U.S. components that support U.S. space exploration goals and components provided by foreign partners, planned by the end of the decade.

3.1 NASA shall focus use of the space shuttle to complete assembly of the International Space Station.

3.2 NASA shall focus U.S. International Space Station research and technology on supporting space exploration goals.

3.3 NASA shall separate transportation of crew and cargo to the International Space Station to the maximum extent practical.

NASA shall pursue opportunities for international participation to support U.S. space exploration goals.

NASA shall pursue commercial opportunities for providing transportation and other services supporting the International Space Station and exploration missions beyond low earth orbit.

NASA shall identify and implement opportunities within missions for the specific purposes of inspiring the Nation.

NASA Also baselined the Executive Council Approved Level 1 Objectives—derived from the Level Zero Requirements:

(1) NASA shall develop and demonstrate power generation, propulsion, life support, and other key capabilities required to support more distant, more capable, and/or longer duration human and robotic exploration of Mars and other destinations.

(1.1) Starting no later than 2008, NASA shall initiate a series of robotic missions to the Moon to prepare for and support future human exploration activities.

(1.2) NASA shall conduct the first extended human expedition to the lunar surface as early as 2015, but no later than the year 2020, in preparation for human exploration of Mars and other destinations.

(1.3) NASA shall explore Jupiter's moons, asteroids, and other bodies to search for evidence of life, to understand the history of the solar system, and to search for resources.

(2) NASA shall separate crew from cargo for launches of exploration missions to the maximum extent practical.

(3) NASA shall conduct the initial test flight for the crew exploration vehicle before the end of the decade in order to provide an operational capability to support human exploration missions no later than 2014.

(4) NASA shall retire the Space Shuttle as soon as assembly of the International Space Station is completed, planned for the end of this decade.

(5) NASA shall conduct International Space Station activities in a manner consistent with U.S. obligations contained in the agreements between the United States and other partners in the International Space Station.

(6) NASA shall acquire cargo transportation as soon as practical and affordable to support missions to and from the International Space Station.

NASA shall acquire crew transportation to and from the International Space Station, as required, after the Space Shuttle is retired from service.

Going from zero to a program

The Level Zero requirements are but the first step in a long process. Consider that the implementation of this new space policy was going to require a large number of things to go just right.

When asked how he was scrubbing new NASA exploration plans in this regard Steidle said that he had an independent cost assessment underway, that he had the Systems Management Office at NASA Langley Research Center doing a technology assessment (due at the end of June 2004), and that he had the Aerospace Corporation looking at alternative approaches.

Of course, Steidle was not walking into a blank sheet of paper—he had a number of existing programs and 'legacy systems' that he needed to either blend into his program or discontinue. In other words, "what to keep – and what to stop."

Steidle showed a series of charts which documented the large number of trade studies he had underway across the agency. These trades looked at the full breadth of what the agency was doing—or was planning on doing. The idea was to have a snapshot of where the agency stood as the new space policy was being implemented.

As part of the process of seeing what NASA had, what it needed, and where it needed to go, Steidle's organization did a survey of space and aeronautics programs. Some programs were continued, others (such as the RS-84 rocket engine and the X-43c) were brought to an end—the criteria being if they fit in where NASA needed to go.

As far as in-space propulsion (upper stages or 'space tugs' capable of moving things around in space) his organization reviewed NASA's Demonstration of Autonomous Rendezvous Technology (DART) program, the Department of Defense's Orbital Express Program, and another program which he declined to talk about (for either national security and/or proprietary reasons). All of these programs involve the use of autonomous means to navigate and dock in space. Steidle's interest was to understand where the nation was headed with regard to robotic rendezvous and docking technology. While some viewed the cancellation of the Space Shuttle SM4 Hubble Servicing Mission as a looming problem, Steidle viewed it as an opportunity and was looking to support NASA's Office of Space Science as it moved through various options that would involved robotic servings and life extension of Hubble.

But how do you transform a presidential speech into spacecraft? At one point in June, presentation to the NASA Advisory Council Steidle showed a chart that described his 'strategy-to-tasks-to-technology' process. This chart was meant to show how Steidle took the president's space policy, using the Level Zero requirements as a departure point and fed those requirements through a defined series of trade studies, development of system requirement documents, modeling and simulation, and an investment plan, so as to transform its words into engineering and programmatic guidance. This guidance would then lead to the specific programs needed to accomplish the policy.

As a test of the veracity of his system, Steidle said he had run the process completely in reverse allowing specific programs to be traced directly back to direction given by the president. As such, if Steidle's charts are indeed accurate, to use a phrase popular with biologists, 'form follows function' in the methodology with which NASA plans to proceed in the implementation of the president's policy.

On paper that is.

But much more is needed in order to put the program into place. Big pictures need to be evaluated and distilled into marching orders. The Level Zero requirements now needed to be 'flowed down' to additional levels with each level having an increased level of specificity. Eventually, as the process continues, you get to the specifications for what color to paint which part. But that was a long way off.

Getting everyone involved

At the core of Steidle's JSF-inspired plan, one thing that needed to be done early and often was to involve the contractor community—as well as other 'stakeholders' and possible participants—in the collection of ideas and the proposing of ways to carry out the space vision. To do this, Steidle would hold a number of 'Industry Days' and pre-solicitation and pre-bidder conferences for the contractor community, to explain progress being made in implementing the new space policy, to provide updates on his plans, accomplishments, challenges, and opportunities.

On 21 April 2004 OExS released a Request for Information (RFI) which explained NASA's acquisition strategy and solicited ideas (in the form of white papers) in a variety of areas, specifically to address, "Initial challenges facing Project Constellation and Project Prometheus in general, and the Crew Exploration Vehicle (CEV) in particular." Some 1,100 responses to this RFI were received by its due date a month later.

A month later, on 21 May 2004 OExS released an Intramural Call For Proposals (ICP) for Human & Robotic Technology. It generated 1,300 Notices of Intent (NOI) by the due date three weeks later.

Clearly there was considerable interest in what NASA's Office of Exploration Systems was up to. These events—industry consultation, Level Zero requirements, trade studies, etc.—led up to the release of three NASA Broad Agency Announcements (BAA) (a common DoD approach but one rarely used at NASA) and a RFP for Project Constellation that provided the first clear definition of the first steps NASA was going to make to implement the president's policy.

At least three such studies are planned, a BAA to address the whole CEV concept; a BAA to study advanced science needs, and a BAA to assess a development path to mature the technology that the Constellation project would require. "We will also extend a major outreach to the scientific community," says Steidle, to craft research programs and science priorities that the lunar-walking astronaut and robotic expeditions will follow. Other BAAs may be added later, or these planned studies changed as a result of industry input. At this early stage, little is fixed in stone, other than the president's directives.

The first BAA, for 'Project Constellation Exploration & Refinement' was released in late June 2004. Two 'concept areas' were covered by this BAA—Concept Area 1: Preliminary Concepts for Human Lunar Exploration and Concept Area 2: Crew Exploration Vehicle concepts. Offerors could submit proposals to Concept area 1 alone or to both areas together. Proposals were due on 16 July 2004 and awards were expected to be made at the beginning of September 2004.

The Second BAA covering 'Technology Maturation and ASTP'—specifically 'systems of systems' technologies was scheduled for release in early July 2004 with contract awards made in September 2004. The Third BAA, also covering 'Technology Maturation and ASTP'—specifically 'Gap Filler technologies' would be issued in late 2004 and awards made in Spring 2005.

In order to force a level of seriousness on proposers, NASA would require that a Notice of Intent be submitted for the two Technology Maturation and ASTP BAAs which would take the form of a white paper. Only once NASA had read the white paper and deemed the content merit-worthy would it request a full fledged proposal from the offeror.

Defining the CEV

In January 2005 a Request for Proposals (RFP) would be issued for the Crew Exploration Vehicle Level 1. An award is planned for May 2005. CEV Spiral 1 will involve the design and development of the first generation of CEV capable of carrying astronauts only into Earth's orbit. Spiral 1's objective calls for sending the first crewed CEV into space no later than 2014, preceded by unpiloted flight tests in 2011 and the first test flight of a stripped-down prototype in 2008.

According to Michael Hecker, OExS Deputy AA for Development Programs, CEV testing would center around choosing two contractor teams that would each launch a prototype of their CEV design, along with a separate booster rocket, into space during 2008. "This is not a fly off," Hecker said, because the test versions would only contain about 30 percent of the systems of a fully developed craft. "The purpose of these demonstrations is risk reduction," added Garry Lyles, Deputy Division Director for the Constellation project. "We'll use the results of the 2008 demonstrations to help us in the source selection of the CEV," Lyles explained.

As for what issues Spiral 1 would address Hecker said the planned issues to be resolved would be done so as to "provide a concept of a CEV—what does it look like? What is its mold line (shape) and subsystems?" In the development plan of the ship, Hecker said that NASA would expect the contractor teams to 'identify the objectives' to be met when the prototype CEVs fly to space in 2008. "What is the initial allocation of (the CEVs) functionality ?" he explained. In other words, how would it perform its minimal mission: Earth orbital flight?

At the end of that development cycle, the resulting design would fold into Spiral 2, which would be a CEV version fully capable of supporting astronauts on their trip from Earth's orbit to the moon and down to the lunar surface. Spiral 2, by definition, would clearly be a more complex CEV. Given Bush's stated objective of crewed lunar landings beginning as early as 2015 but no later than 2020, the lunar CEVs would be not just a single ship, but a series of 'technical solutions' that accomplish Bush's lunar element. The Moon-bound variants would come with a whole series of equipment, according to Steidle, that supports humans on the Moon for extended stays.

Garry Lyles adds that the lunar CEVs should also have systems and elements that could migrate to their Mars-bound CEV variants, when that version began development on a new spiral in future years. Such a migration is referred to by the OExS crowd as 'extensibility' i.e. can the capabilities in one version of CEV hardware be extended forward for future uses—and future needs are reflected in earlier versions. In the software world, such capabilities would be referred to as 'forward' and 'backward compatibility'. Steidle's team was clearly thinking ahead.

Such a future Spiral, not yet in the detailed planning stage like Spirals 1 and 2, would be a CEV version fully capable of interplanetary flights to Mars. Steidle pointed out that President Bush's vision called for human voyages to other destinations in the solar system, such as asteroid rendezvous, or someday a truly bold mission to the moons of Jupiter. "It's not aimed at one destination," Bush said.

What about management of the operational fleets of CEVs and their launch systems? Would NASA seek to recreate the type of Space Flight Operations Contract (SFOC) that is used for the Space Shuttle fleet? "We are open to whatever the industry brings to us," Steidle explains. And what technologies does NASA seek to mature? The CEVs will have all-new heat shields, made of new materials that didn't exist when the space shuttles were designed 30 years ago; even technologies that didn't exist when the X-33 was under development in the mid 1990's. While no decision has been reached as to whether the CEVs will be expendable or reusable, reusability as a favored feature seems likely. Even the technology of landing systems is still undetermined. Will the CEV return to Earth by parachute, on landing gear, by rocket thrust—or some combination thereof?

Beyond the design and technology of the CEV is the issue of how they will find their way to the Moon. Mission modes are also under trade study review. Will the CEVs fly from Earth directly to the Moon, or will the expedition fleet be assembled in Earth orbit and then embark for cislunar space? No pathway has been selected yet as this book went to press in mid-summer 2004.

CEVs and the ISS

The task before Steidle and his team is a multi-pronged effort to get the first requests out to industry on how to develop the Constellation spaceships. The spacecraft will come in slightly different versions, like different models of a car. Some versions will come with hardware that would allow it to serve as a base of operations on the Moon's surface. Yet other variants will be capable of leaving the Earth-Moon system behind and head out for asteroids, or Mars, or possibly more distant destinations.

When it was originally proposed, the Orbital Space Plane was to initially serve as a Crew Return Vehicle (CRV) for the space station—a function carried out in an 'interim' fashion by Russian Soyuz spacecraft. Eventually like the OSP it would also take on the Crew Transport Vehicle (CTV) function—now provided by the Russian Soyuz and the shuttle fleet. Although Sean O'Keefe would make constant reference to lessons learned in the short lived OSP program, some had originally interpreted the CEV, when it was announced, as something that would do everything the OSP would have done—plus go to the Moon. It would soon become clear that this was not the case.

The space shuttle fleet is scheduled for retirement around 2010 once assembly of the ISS is completed. After that transport of crew, including Americans, would be provided by Soyuz via a yet to be determined agreement with Russia. Whether this was a permanent solution is uncertain. The ISS would now be used to do exploration-related research on humans. The first CEV demo flight is planned for 2008. The first crewed CEV would not be launched until 2014; around the time that attention would start to shift toward sending humans to the Moon.

Following release of the space policy, NASA has been increasingly less precise as to what the policy entails in terms of U.S. participation in the ISS program. Given the tight budgets involved it is not likely that both a full blown ISS research program and a developing program to send humans to the Moon would be affordable. Nor is the ISS, at an inclination of 51 degrees, a likely departure point for missions to the Moon. When asked if the CEV would visit the ISS, Steidle said in June 2004, "We don't know that yet." As such the prospect was apparently quite real that America would deliberately decide not to develop an independent way to visit the ISS—something it had spent tens of billions of dollars and immense political capital on.

Prizes

Perhaps one of the more unusual aspects of how Craig Steidle approached stimulating interest and new ideas was the 'Centennial Challenges' program he initiated. This program was described officially as being "a program of contests in which NASA will establish cash awards to stimulate innovation and competition in technical areas of interest to Civil Space and Aeronautics." The program was organized in what NASA saw as a low risk fashion so as to encourage innovation in ways that standard federal procurement cannot; enrich NASA research by reaching new communities; help address technology pitfalls; promote returns that outweigh the investment; and educate, inspire and motivate the public."

The Centennial Challenges program sought innovation in "Revolutionary advances in fundamental technologies; Breakthrough robotic capabilities; and very low cost space missions." As initially created, the prizes were open to all U.S. citizens not directly employed by NASA or the federal government. That opened the door to non-profits, schools and universities, and it was hoped, especially entrepreneurs. The idea being to stimulate the private sector to develop innovative technologies that would be at the center of the new project. The prizes offered in FY 2004 were limited to $250,000 although future prizes could be much larger.

Prizes have certainly had an effect on how some innovative aerospace capabilities and exciting events have come into being. Charles Lindbergh's solo flight across the Atlantic was spurred by the $25,000 prize offered by Raymond Orteig which would be awarded to the first person to fly nonstop between New York and Paris. Paul MacCready won the $100,000 Kremer prize when his ultra-lightweight human-powered Gossamer condor met the competition's conditions. MacCready won another Kremer prize for $200,000 when his Gossamer Albatross was powered and piloted by a human as it flew across the English Channel.

Of course, the most recent and relevant example was the $10 million Ansari X-prize which spawned dozens of teams to attempt to win a prize by privately financing, building and launching a space vehicle which is able to carry three people to an altitude of 100 kilometers (62.5 miles), returning safely to Earth, and then repeating the same feat—with the same vehicle—within 2 weeks.

On 21 June 2004, pilot Mike Melvill climbed into the cockpit of SpaceShipOne. An hour or so after take off his plane and its mother ship reached launch altitude. Melvill was released and fired his plane's rocket engine for a ride into history. By the time his rocket engine had quit, and his plane had reached the high point in its flight, he looked down at the Earth from 328,491 feet. He was in outer space. The first person to make it to space on a privately funded spacecraft. Upon landing it was discovered that a number of rather serious problems had occurred during the flight—problems which would take some time to correct. Melvill hadn't won the X-prize for SpaceShipOne inventor Burt Rutan, yet—but he certainly reduced the doubt that this would soon happen.

Rutan and his prime financier Paul Allen put more than the amount of the expected prize into what they could win ($20 million invested for a $10 million prize). As such, they claim that the prize was not the prime motivating factor in their efforts. However, they are trying to win it, and the X-prize clearly spurred many involved to attempt this feat in the first place. Lopsided investment/prize ratio not withstanding, Rutan says he will happily cash the prize check.

Moving forth boldly

An indication of the 'tone' that NASA's Office of Exploration Systems would take could be seen in a logo which Craig Steidle first showed publicly at a 7 May 2004 Presolicitation conference. The logo was designed by Michael Okuda, Scenic Art Supervisor and Technical Consultant for Star Trek at Paramount Studios. Within its design are three overlapping worlds – Earth, Moon, and Mars, and a NASA vector-like streak arching past all of them to the distant stars. The word 'Exploration' is on the top and 'NASA' is on the bottom. The Latin motto 'Audentes fortuna juvat' is on the left and again its English translation 'Fortune favors the Bold' on the right.

Variously translated as 'Fortune assists the bold' or 'Fortune befriends the bold' the quote comes from Virgil's Aeneid – the national epic of Rome written between 30 to 19 B.C. To be specific, the quote comes from Book 10 of the Aeneid which describes a pivotal battle during the last days of Troy:

> *Be mindful of the race from whence you came,*
> *And emulate in arms your fathers' fame.*
> *Now take the time, while stagg'ring yet they stand*
> *With feet unfirm, and prepossess the strand:*
> *Fortune befriends the bold." Nor more he said,*
> *But balanc'd whom to leave, and whom to lead;*
> *Then these elects, the landing to prevent;*
> *And those he leaves, to keep the city pent.*

Given the challenges that Steidle faces not only with the technology, but also with changing the way NASA does business, the phrase is certainly appropriate. However not everyone at NASA thinks so.

NASA can be very sensitive to things like logos, since they can, like other symbols and phrases, send messages, that people who are looking to raise issues, can use in ways the agency might not like. As such, an immense amount of time goes into crafting the agency's public face—and the default choice at the end of that laborious, byzantine process, is invariably conservative.

While Steidle's organization liked the jaunty 'in your face' flavor of the phrase, it did not sit well in the mind of NASA Assistant Administrator for Public Affairs Glenn Mahone. Mahone is a public relations veteran and does not make decisions or adopt a stance based simply on whims. But instincts regarding the best face for the agency and an emerging, more energetic face that represents what the agency wants to become can often be at odds with such conservatism.

Therein began a low level dispute as to whether this phrase represented NASA as an agency (nothing like it was in the president's statement) or whether it was just something that the Office of Exploration Systems would use. Since this logo, and its wording ,was going to be used to represent at least one part of the implementation of the president's policy implementation at NASA—arguably the most expansive—by default it represented the agency as a whole. The compromise reached (as the logo was eventually animated for TV use) was that Steidle would use the phrase, NASA, as an agency would not.

Even as new efforts within NASA were poised and willing to take a leap out into the solar system, other forces within the agency sought to pause at the cusp.

There was no one at NASA that seemed to mind that the Mission Operations Directorate (MOD) at NASA Johnson Space Center had a long-used logo which has been revised visually since the earliest days of human space flight but has always said 'Res Gesta Per Excellentiam' (Achieve through Excellence). Nor did they seem to mind the logo developed after the Columbia accident which is found on the back of the CAIB report. This logo (also designed by Michael Okuda along with NASA's Bill Foster) shows design elements from the STS-107 Columbia, STS-51-L Challenger, and Apollo 1 logos with the motto 'Ad Astra Per Aspera – Semper Exploro' on the top 'to the stars through adversity – always exploring'.

Sometimes it seems that NASA's organizations aren't at all shy about what inspires them, but their head office is. The agency is going to have to confront this schism sooner rather than later if it is going to be up to the task that has been set before it. If it cannot agree on something as simple as a logo, how will it ever get beyond Earth orbit?

How to officially say "vision"

Soon a discussion began about how to refer to the vision itself. The official way to refer to it was eventually captured as VSE – Vision for Space Exploration. This sprang from a lot of internal confusion. On 12 May 2004 an email message made its way around NASA. It said

"Please note the following message from the Office of Public Affairs:

From now on, per Glenn Mahone, when we talk about or write about the new space exploration vision that President Bush put out in January, please refer to it as:

The Vision for Space Exploration .

It's not the president's new vision, it's not NASA's new vision... it's just the Vision for Space Exploration. Since it's a new initiative, we're treating it as a proper noun, so the V, S and E get capitalized. This isn't something we should actively tell the media about. Just start casually using it this way from now on."

This is typical of NASA's odd obsession with the way words are used. NASA people love to 'action' things; 'Field Centers' is capitalized; and everyone writes everything in uppercase PowerPoint snippets.

Meanwhile, as NASA headquarters fussed with logos and mottos, the rest of the agency seemed to be dragging their feet about even mentioning the new policy—whatever HQ wanted to call it. As of the six-month anniversary of the president's announcement, only three NASA field centers – Ames, Stennis and Johnson had links from their own main website to the official Vision for Space Exploration information at the NASA home page. All other field centers apparently saw no need to make mention. Then again, with only two exceptions, none of the field centers made links to other field centers readily available, and links to NASA's home page were often a bit counter-intuitive, so this was not exactly a new type of behavior. This situation was only resolved after a posting appeared on NASA Watch – and after Sean O'Keefe ordered the field centers to place links back to the agency's central page describing the 'Vision for Space Exploration'.

As the Aldridge report landed on the president's desk, Craig Steidle and his merry band of reformers were outlining the implementation of the president's space vision at a frenetic pace at NASA headquarters. But the rest of the agency seemed to be only reluctantly embracing the vision. Clearly some changes were needed in how NASA itself was set up to do such things. By no coincidence that is exactly what the Aldridge Commission focused on.

*   *   *   *   *

12

THE PATH TO TRANSFORMATION

Remake NASA

As he announced his new space vision, George Bush also announced the beginning of the process of seeking external input and guidance as this vision was to be transformed into reality. When he made his speech, NASA had only the general conceptual framework and large scale budget issues resolved. Exactly what sort of vehicles would be needed, when and where they would go, and what they would do, was reflected only in a few hundred words.

For NASA to have spent too much time working on the implementation of the plan before it had been approved—which happened on 19 December 2003, would have been somewhat presumptuous. To have started to work on it after the internal agreement to move ahead would have preempted the formal announcement of the policy in the manner the president had chosen. In addition, most of the people who'd do the implementing were burning 'use or lose' vacation time at the end of the year.

Once the policy was set forth and implementation was to begin, it was to be done in a very transparent fashion. Part of that transparency was the creation of the 'President's Commission on Implementation of U.S. Space Exploration Policy'. Soon known as the 'President's Commission on Moon, Mars, and Beyond', and eventually as 'The Aldridge Commission', this effort would be headed by former Air Force Secretary Edward C. 'Pete' Aldridge.

The commission would be tightly focused on reviewing, not supplementing, the space vision plan, and reporting back to Bush by early summer. It would not be debating whether to go to one destination or another. Bush had already made that decision. Rather, this report was about taking a vision as outlined in a speech and transforming it into an actual program.

This would also be a swift process. There would be no foot-dragging or ponderous reports. The Commission would exist for no more than 180 days and was chartered to provide its final report to the president within 120 days of its first meeting.

The Commission

According to the commission's website: "Pete Aldridge is a 42-year veteran of aerospace technology leadership, serving the nation for more than 18 years in the Department of Defense, most recently as the Under-Secretary for Acquisition, Technology, and Logistics. His DoD career began as an operations research analyst. Later he became Under-Secretary and then Secretary of the Air Force under President Reagan."

Aldridge also worked at the Aerospace Corp. trained to become a military astronaut (but never flew), and was on a rumored short list to succeed Dan Goldin to become Administrator. Aldridge also underwent training in the mid 1980's with the prospect (never materialized) of flying as a Department of Defense Payload Specialist. Aldridge was a consummate Washington insider who had managed to be involved in a rather broad range of high technology programs and systems—a lot of it hands-on. It would not take him much time to get up to speed on the topic at hand.

Aldridge's panel members represented a mix of science, government, military, educational, and commercial backgrounds.

Carleton (Carly) Fiorina is chairwoman and chief executive officer of Hewlett Packard. Outspoken about nearly everything, she sparked controversy at HP with the merger with Compaq—rising the ire of the Packard family in the process.

Michael Jackson was senior vice president for AECOM Technology Corporation and served as Deputy Secretary of the Department of Transportation. Before that Jackson was chief operating officer of Lockheed Martin Corp.'s intelligent-transportation systems unit.

Laurie Leshin was the director of Arizona State University's Center for Meteorite Studies and is involved with a number of past, present, and future space missions.

General Lester L. Lyles had a 35 year career in the U.S. Air Force for more than 35 years retiring as a 4-star general in command of the Air Force Materiel Command.

Paul Spudis was a planetary scientist at the Johns Hopkins University Applied Physics Laboratory whose specialty is the lunar geology—and has studied Mars, Mercury, and other worlds as well.

Neil Tyson was an astrophysicist and Director of the Hayden Planetarium in New York City. Tyson also served on the Commission on the Future of the U.S. Aerospace Industry.

Robert Walker was chairman and chief executive officer of The Wexler & Walker Public Policy Associates and was a Republican Congressman from Pennsylvania from 1977 to 1997. He was Chairman of the House Science and Technology Committee and was the Chair of the Commission on the Future of the U.S. Aerospace Industry.

Maria Zuber was a professor of Geophysics and Planetary Sciences at the Massachusetts Institute of Technology and has been involved in more than half a dozen NASA planetary missions aimed at mapping the Moon, Mars, Mercury, and several asteroids.

Full transparency

The Commission operated under the Federal Advisory Committee Act, and, as such, was required to do most of its deliberations in public. In order to get input from so-called stakeholders (NASA, commercial, scientific, and other interests) and the general public, the commission held 5 field hearings starting in February in Washington DC, followed by hearings in Ohio and Georgia in March; California in April, and finally in New York in May. Full transcripts of all that was said, copies of presentations made, and video of the hearings were posted promptly on the Commission's website. In addition, all hearings were broadcast live on NASA Television.

The panel heard from former advisory commission chairs such as Thomas Stafford and Norman Augustine, Sean O'Keefe, representatives from industry and science, professional and space advocacy groups and the general public.

At the end of each hearing, members of the audience were selected at random from a larger list and allowed to make short presentations to the commission. In addition, more than 6,000 suggestions were received via the commission's website. Their task was hurried, but they managed to plow through a lot of territory and give a large number of people a chance to have their say.

Other such commissions have longer periods in which to operate—since they only report back on what might be done. This commission was given a strict sunset clause and was expected to provide guidance to a series of policy directives the president had already implemented. They did not have the luxury to sit back and ponder.

Moreover, little of what finally appeared in the Commission's final report was all that much of a surprise since all of its public hearings were widely covered, and the presentations and Commissioner's questions clearly allowed major issues and concerns to become readily apparent.

Delayed release

The commission had been scheduled to release its report in Washington, DC on 9 June 2004. However, with the death of Ronald Reagan, and the vast number of mourners about to descend on Washington this event was rescheduled The new date was 16 June 2004. Since the event had more or less been primed to happen on a certain day, review copies of the final report had already begun to be distributed. Adding in an additional week of time allowed the report to spread widely and within a few days of its original planned release—and still days away from its formal release—everyone already knew what was in it. No one was really surprised.

The report itself was short. Some thought it was too short and superficial and that some additional work was required. Others, most notable NASA employees complained when they were asked to provide their views within several days saying that it would take much longer than that to read it. No one is ever satisfied with an advisory committee report.

NASA's initial response was neither to embrace or dismiss this recommendation. However Sean O'Keefe has indicated that, "The recommendations released today by the commission will influence our work for years to come and will help guide us through a transformation of NASA." O'Keefe saw the report as being just long enough to provide him some guidance but not so long and detailed as to limit the way he implemented its recommendations.

Media response was somewhat ho-hum. An opinion piece in the New York Times was somewhat ambivalent about the report and observed of the Commission's recommendations, "Congress will need to take these mostly half-baked proposals and cook them more thoroughly." Local newspapers seemed to focus on one thing; a controversial suggestion that NASA's field centers be converted into a different sort of federal research laboratory.

Other reporters focused on the things that their readers were most concerned with. An article in Hampton Virginia's Daily Press, a newspaper read by NASA Langley Research Center employees, ignored the broader aspects of the space exploration vision and focused on possible job losses. Similar articles appeared in newspapers read in communities hosting other NASA field centers.

While the president's space policy announcement had garnered worldwide attention in January for almost a week, news surrounding the release of the Aldridge Commission's report was already fading after one news cycle.

Aldridge does the Hill

After the report was released, Pete Aldridge appeared before Congress. On 17 June 2004 he appeared before the Senate Commerce, Science, and Transportation Committee. Chairman Sam Brownback (R-KS) referred to the report as, "a thoughtful and compelling roadmap for our future exploration activities." He also pledged to help implement the president's policy by introducing a NASA authorization bill designed to cover the next five years. Brownback had made such a pledge early after the Columbia accident in 2003.

Sen. Bill Nelson (D-FL) repeated earlier concerns about what he saw to be insufficient budgetary commitment for the overall plan and was happy to see that the Commission had endorsed the idea of having a presidential advisory commission to coordinate space policy—another issue he had long spoken of favorably.

O'Keefe responds

On 24 June 2004, the day after Pete Aldridge made his appearance before the House Science Committee, the details of O'Keefe's plan to start the transformation of the agency along the lines suggested by the Aldridge Commission's report were ready for release.

The media and the agency were alerted to an announcement several days before. This was followed by additional updates sent out to all employees. Meanwhile, the Office of Science and Technology Policy (OSTP) was working hard to drum up attendees for an event just before O'Keefe would hold a formal press conference. The White House was eager to show that the president was fully behind this event. The press office at OSTP sent out personal invitations to industry representatives all over Washington inviting them to a briefing at 2:00 p.m. in room 450 of the Eisenhower Building. The larger companies could bring as many as 5 people. The intent was to have a standing-room only crowd in a room which sat 150. Craig Steidle would brief the attendees on the transformation plan.

He was joined afterward by Andrew Card, the White House Chief of Staff. Card told the audience that Bush was still strongly behind his space agenda. He also entertained the group with stories from Bush's first days in the presidency.

At 3:00 p.m. a press conference was held at NASA headquarters in the Webb Auditorium on the same stage where President Bush had spoken only six months earlier. The overall motif was 'NASA Inc.'—treating the agency as a corporate entity with disparate efforts moved from a diffuse management scheme (which was in place) toward a more central, 'corporate' approach.

Space Exploration Steering Council

Among the recommendations was a call for the president to establish "a permanent Space Exploration Steering Council, reporting to the president, with representatives of all appropriate federal agencies, and chaired by the vice president or such other senior White House executive that the president may designate. The council shall be empowered to develop policies and coordinate work by its agencies to share technologies, facilities, and talent with NASA to support the national space exploration vision."

As with all of the compact recommendations made by the Aldridge commission, NASA, with the counsel of others, was going to look at how this might be done— and if there were alternate ways to do so. However, given the constant call for a body similar to the old National Space Council, it is hard to imagine how NASA and the White House will be able to avoid creating the Space Exploration Steering Council unless they choose to disregard this recommendation.

This notion of a high level advisory body for space policy is not new. On 5 November 2003, in response to concerns raised in the Fall of 2003 as the Bush space policy was being formulated behind closed doors, Sen. Hollings (D-SC) and others cosponsored S.1821 the 'National Space Commission Act'. This Act, if enacted, would direct that this commission have "12 Members, who shall be appointed by the president by and with the advice and consent of the Senate." Which would "provide advice and counsel to the president and the Congress of the United States on matters related to the future development and use of space," among other things. The bill never added any co-sponsors to the original six Democratic senators - Sen. Breaux (D-LA), Sen. Dorgan (D-ND), Sen. Inouye (D-HI), Sen. Kerry (D-MA), Sen. Lautenberg (D-NJ), and Sen. Rockefeller (D-WV) who lent their support when the bill was introduced. No action was taken on the bill by the Senate after its introduction.

The creation of the Aldridge Commission at the same time that Bush announced his new space policy took some of the lingering momentum out of a need for a presidential-appointed body to look over space policy—only to then come back and call for much the same thing that Hollings et al had called for in November 2003.

Elevating the Administrator's stature.

At one point in the process, Aldridge's commission had forwarded a number of things it was thinking of recommending to the White House. One idea was to elevate the nature of the position of administrator of the agency. The suggestion was that the position be modeled after that used at the FBI where a director was appointed to a term—one that would allow the position to become independent of political whims. Aldridge's Commission suggested a term of 10 years.

The original plan had been to roll out the report at the U.S. Chamber of Commerce hosted by the Space Enterprise Council. At one point Pete Aldridge presented this 10 year appointment concept to Thomas Donohue, President and CEO of the Chamber. Donohue said that the Chamber would support that idea if it were in the Commission's final report.

The White House did not have any problem with the idea. Sean O'Keefe and others were consulted as well. A suggestion was made that someone had already served in that capacity—for that length of time—and wondered if that was the model that the Commission really sought to emulate.

The recommendation never made its way into the final report. The legacy of Dan Goldin had struck again.

The Private Sector

The commission observed, "NASA's relationship to the private sector, its organizational structure, business culture, and management processes - all largely inherited from the Apollo era - must be decisively transformed to implement the new, multi-decadal space exploration vision." To fix this situation, the commission recommended that NASA "recognize and implement a far larger presence of private industry in space operations with the specific goal of allowing private industry to assume the primary role of providing services to NASA, and most immediately in accessing low-Earth orbit. In NASA decisions, the preferred choice for operational activities must be competitively awarded contracts with private and non-profit organizations and NASA's role must be limited to only those areas where there is irrefutable demonstration that only government can perform the proposed activity."

These sentiments are certainly not new. In recommending that NASA procure services from the private sector, the Commission was recommending that NASA continue to do what it has done since its inception i.e. procure goods and services from the private sector. One might have a valid point that some things that the agency does itself 'in-house' should also be procured commercially, but the implication from the observation made by the Commission lightly suggests that NASA doesn't prefer to use commercial vendors.

The issue of how much commercial solutions NASA uses is not a new one. On 28 October 1998, President Clinton signed the Commercial Space Act of 1998 creating Public Law 105-303 which calls for a number of policy changes in the way that the government views commercial solutions to things that the government needs to accomplish. With regard to launch services the Act says, "Except as otherwise provided in this section, the Federal Government shall acquire space transportation services from United States commercial providers whenever such services are required in the course of its activities. To the maximum extent practicable, the Federal Government shall plan missions to accommodate the space transportation services capabilities of United States commercial providers." Exceptions include payloads that are only capable of being launched by the space shuttle, where national security concerns arise, or when scientific requirements might be compromised." The act also called for studies on how to privatize the space shuttle fleet, and third party liability issues.

In addition, the Act specified, "The Administrator shall, to the extent possible and while satisfying the scientific or educational requirements of the National Aeronautics and Space Administration, and where appropriate, of other Federal agencies and scientific researchers, acquire, where cost effective, space science data from a commercial provider." Similar language was included with, "Space-based and airborne Earth remote sensing data, services, distribution, and applications from a commercial provider."

While NASA had made some partial progress in adhering to this Act, many would say that the agency had not really seen the need to pick up the pace, and that it often used exceptions in the Act's language to stall further compliance with the Act's overall intent which was to move NASA and other agencies with space needs away from in-house or government-controlled solutions.

What was being developed within the Office of Exploration Systems suggested, however, that some serious reconsideration was being given to the way that NASA involved the private sector. Whether this would extend out to the entire agency was another matter all together.

Echoes of Apollo

The suggestion that the agency's "organizational structure, business culture, and management processes - all largely inherited from the Apollo era" is also problematical. To be certain, the same Apollo-era field centers still exist, many of the iconic buildings from that time (and before) are still in use, and the space shuttle, which was inspired during the Apollo era is still America's prime mode of putting humans into space. But much of what the agency does has undergone an overlapping but regular cycle of upgrades. There is nothing whatsoever Apollo-like in the way that twin rovers are managed as they drive across Mars. The software designers were not even alive during Apollo. But the mission is still managed from a control room located inside a large sprawling NASA center. With today's telecommunications capability you could easily do this almost anywhere.

One could certainly argue that the agency should move faster and be more creative, but a more accurate portrayal of how NASA is configured would be to say that it is a hodge-podge of things from both yesteryear and an hour ago—some of it being used to develop tomorrow's technology just down the hallway. What should really be done is to follow the biologist's mantra of 'form equals function'. If one knows how the agency needs to function, the solution is to alter its form so as to be able function properly in accomplishing the tasks it has been handed. NASA's tendency is usually to do the opposite; with a reluctance in many (but not all) sectors to change. That is probably what the commission was really trying to highlight. Indeed it recommended that, "NASA be transformed to become more focused and effectively integrated to implement the national space exploration vision, with a structure that affixes clear authority and accountability." Indeed, that is what O'Keefe intended to do.

Retool the agency

The Aldridge Commission observed, "NASA's organization chart is not wired for success. The first task is to realign the NASA Headquarters organizations to support the long-term vision. There are currently too many mission-focused enterprises and the mission support functions are excessively diffuse." It then recommended that NASA and Congress "work with NASA to create 3 new NASA organizations: a technical advisory board that would give the Administrator and NASA leadership independent and responsive advice on technology and risk mitigation plans; an independent cost estimating organization to ensure cost realism and accuracy; and a research and technology organization that sponsors high risk/high payoff technology advancement while tolerating periodic failures."

This recommendation (and others) would require for some substantial changes within the agency. O'Keefe had already seen this train coming and had a transformation plan already well defined when the Aldridge Commission's report was released. By virtue of having done so much in public it was straightforward for O'Keefe to see where they were going.

On June 24th Sean O'Keefe began the process of transforming the Agency. This process began at NASA Headquarters. O'Keefe framed the news he would unveil by referring to two things he had heard since he had arrived that affected his decisions. First was the common plea, "We need a strategic direction. We need a focus." "We got that on January 14th," O'Keefe said. The second thing he had heard was the prevailing logic that the agency was 'one accident away from going out of business'. "This was asserted as if there was no debate." O'Keefe said. "Guess what: it did not happen!"

O'Keefe sought to focus all of NASA on its core values and to create a clear roadmap of where the agency was going to go; restructure and increase the integration of how the agency works; and enhance the level of accountability across the entire agency.

Specific changes

As for the details: one would be the use of the term 'codes' to denote various offices. This practice had its origins in the alphabet system of routing mail to various offices and grew into a parallel language that described organizations, without using their names. This was, of course, a feature at all field centers. Some used letters (all in a different syntax than headquarters), some used numbers, some used both. O'Keefe was tired of this organizational balkanization and sought to impose a single, easily understandable system across the agency. Indeed he would soon start to joke that anyone using 'code' to describe a part of the agency would have to drop a dollar in a MWR (morale, welfare and recreation) Fund jar by the door.

He also sought to consolidate roles and responsibilities in a way that focused energies and resources along functional lines—not disperse them. Such a dispersal had been Dan Goldin's way of running the agency. In order to avoid having any one person wield too much power (with the exception of George Abbey of course who ran his own fiefdom within Johnson Space Center) Goldin broke up large offices into smaller ones—increasing the number of Associate Administrators every time. Some joked that he might soon need another alphabet with more letters.

By dispersing authority, Goldin was able to manage the agency more easily—since no one person or office could thwart his plans. Despite the somewhat emasculating effect it had on people and organizations Goldin's approach did actually provided a curious sort of stability for the agency. While NASA was no longer known as an energetic center of innovation, it nonetheless managed to retain the status quo rather well—even as Goldin championed seven years of cuts to the agency's overall budget. Alas, this organization Goldin left behind was best suited for keeping the pilot light on—not firing an afterburner and leaping on to new challenges.

O'Keefe would tackle this problem head on. Enterprises and various offices would disappear. Instead, new directorates would be formed. The Space Operations Directorate, which would manage missions being performed in space would be headed by Associate Administrator for Space Flight Bill Readdy. The Exploration Systems Directorate would be run by Craig Steidle. The old Office of Biological and Physical Research (OBPR) would be absorbed into Steidle's directorate. Former OBPR AA Mary Kicza would become the new Associate Deputy Administrator for Systems Integration.

The Science Mission Directorate would be comprised of the old Office of Space Science and the Office of Earth Science and would be run by Goddard Space Flight Center director Al Diaz. Former AA for Space Science Ed Weiler would move to become the new center director at Goddard. Former AA for Earth Science, Ghassem Asrar would become Diaz's Deputy and would also serve as Chief Exploration Scientist. The Aeronautics Directorate would be run by current Office of Aeronautics AA Victor Lebacqz.

The previous NASA Headquarters offices of Public Affairs, External Relations, and Legislative Affairs would now all be overseen by a new Mission Support Directorate headed by an Assistant Administrator who would report directly to O'Keefe.

All other organizations at Headquarters would be tweaked a bit, but would otherwise remain unchanged.

This new NASA headquarters organization would be run by a new Strategic Planning Council (SPC). Gone would be the Executive Council and the Joint Strategic Assessment Committee. (JSAC). The SPC would be chaired by O'Keefe. Its members would include Deputy Administrator Fred Gregory, Bill Readdy, Craig Steidle, Al Diaz, Victor Lebacqz, ADA for Integration Mary Kicza, ADA for Institutions and Asset Management Jim Jennings, and a new ADA for communications.

Ex officio members of the SPC would include NASA Chief of Staff John Schumacher, NASA Comptroller Steve Isakowitz, Chief Exploration Scientist Ghassem Asrar, and Charles Elachi who would now serve as the agency's Director of Strategic Planning while still retaining his JPL center director position.

In addition to these changes, all of the agency's field center Chief Financial Officers (CFO) would now report directly to NASA's CFO Gwen Brown at NASA Headquarters. Previously, all of the center CFOs had reported to their respective center directors. As NASA sought to reform its Byzantine financial management system—one wherein every center had their own way of doing things—it soon became clear that the agency needed to take additional steps such that everyone, everywhere was singing from the same song sheet.

In addition to central coordination of financial management, NASA would also seek to firm up the management of its corporate image—communications and legislative affairs. Among the changes was having the public affairs operations at each center begin to coordinate their activities under the direct management of NASA headquarters.

As the summer progressed, additional personnel changes would be announced which sought to further tighten and control this new 'corporate' style of NASA management with more and more coordination coming to Headquarters—and away from the field centers.

A decade earlier, Dan Goldin had sought to give more authority to the field centers. The net result was what came to be called 'stove piping' where all centers became dueling fiefdoms—all intent on having their own mini-NASA complete with efforts which unabashedly duplicated those resident at other centers. Everyone had a piece of something. Headquarters was often reduced to negotiating with its field centers instead of directing them. Given the way Goldin had diluted the power base at headquarters, the field centers got away with it. O'Keefe now sought to reverse that pendulum and retool the agency as needed to accomplish the challenge that lay ahead.

Convert the field centers

Any time an overhaul of NASA is discussed, the issue of 'what do we do with field centers?' comes up within the first 30 seconds. NASA does its science and engineering at its field centers. Each center has a fierce, independent identity which has grown and matured over the decades. Talk to anyone at the field centers and they will eventually say that they'd just like NASA Headquarters to send money and leave them alone to do their job.

Over the years, as discussed above, the field centers have developed their own collection of programs which they are determined to protect. As such, they have each developed their own support network—one which includes contacts with Congress which can often surpass those performed by NASA Headquarters in terms of sophistication and ability to pass along information. As such, whenever NASA headquarters starts to hint along lines not to the liking of a field center, the message comes back to Headquarters rather fast, from one of the large marble buildings up the street, to 'watch out'.

The net result of this situation is that the agency is often incapable of getting out of its own way. Often times, people marvel at the ability of NASA's workforce to pull off some of the miracles that they still manage to do. If nothing else this serves as a testiment to just how skilled NASA's family is; to be able to excel despite management practices that wouldn't survive for an instant outside in the real world. However, when the management system fails, as was the case with the Columbia accident and all of the email volleys that came to light, NASA's frailties are suddenly exposed for all to see.

The Aldridge Commission recommended that "NASA Centers be reconfigured as Federally Funded Research and Development Centers to enable innovation, to work effectively with the private sector, and to stimulate economic development. The Commission recognizes that certain specific functions should remain under federal management within a reconfigured Center."

NASA's response to this recommendation was, to be uncomfortable to say the least. Pete Aldridge did mention that some discussion had been made for recommending a BRAC (Base Realignment and Closure) function like what the Department of Defense uses—to reexamine field center roles and responsibilities and close a few down if they don't pass muster. Aldridge said this was dismissed because it would have been simply too much to expect NASA to accept.

During Dan Goldin's orchestrated managerial upheavals in the 1990's—Red Team/Blue Team, Zero Based Reviews, etc.—the bogeyman of center closure visited the minds of people at Ames, Lewis (now Glenn), and Langley research centers. Nothing ever came of this, but everyone took notice. Some, most notably Ames, ran with the notion even when not under imminent threat and began to create a research park carved out of its real-estate. Other centers expanded on traditional university and industry alliances. But other than Ames, none of the field centers really saw the need to transform themselves into something new. Besides, Congress would never allow a center to be closed, or so the common logic went.

The problem with this recommendation is that it specifies a one-size fits-all solution to how NASA's scattered field centers are operated. One of its more prominent, JPL, is already managed under such an arrangement—a rather long standing one—with the California Institute of Technology. Indeed some suggestion is made that JPL might serve as a good model in terms of excellence in program management and cost control for other centers to emulate. To be certain JPL-managed missions have revealed an entire solar system to humanity. But JPL also crashed two perfectly good spacecraft into Mars in 1999 because it cut too many corners and bragged about how efficient it was becoming in the process. As far as cost efficiency, there are many in the agency who would suggest that JPL employees are grossly overpaid and much of what JPL does is needlessly gold-plated. This does not mean the JPL model is a bad one to follow, but perhaps JPL should adhere to it as well.

NASA's response to this recommendation was to go off and mount a serious study of what it needed its field center to do and how it might improve their performance. As part of the transformation announced by Sean O'Keefe on 24 June 2004, in response to the Aldridge report, the field centers would find their core missions focused and that they would answer to Headquarters in new ways. JPL, Ames, and Goddard would report to the new Science Missions Directorate. Glenn, Dryden, and Langley would report to the Aeronautics Directorate, and Johnson, Kennedy, Marshall, and Stennis would report to the Space Operations Directorate. Previously

Ames, Dryden, Glenn, and Langley reported to the Aeronautics Office at NASA headquarters as a legacy of how they were originally configured in the early days of the agency. This continued even as a substantial portion of their work had nothing to do with aeronautics; an Apollo era legacy.

Given that the field centers still hold on to all of their previous research portfolios—portfolios which are duplicative and waste the agency's resources—follow-on culls and adjustments wherein these overlaps are reduced or eliminated will be made. Moreover, each field center's roles and responsibilities are certain to be refined and refocused along lines that support the president's space policy.

This will not happen smoothly and is likely to be one of the most daunting challenges O'Keefe will face. With few exceptions, a visit to one or another of NASA's field center's main websites will not yield links to any other field centers. If you can't even get a website to put in a link, and thus an acknowledgement, of the fact that each center is but one of many field centers, you have to wonder what headaches will come with overhauling their core mission and responsibilities.

Spiral development

With regard to the Aldridge Commission's recommendation that "NASA adopt proven personnel and management reforms to implement the national space exploration vision, to include: use of 'system-of-systems' approach; policies of spiral, evolutionary development; reliance upon lead systems integrators; and independent technical and cost assessments," the way that Craig Steidle had already been running the Office of Exploration Systems for several months already embodied all of these recommendations. Indeed, in the eyes of some, Steidle's example may have had the effect of influencing the Aldridge commission to issue this recommendation not so much as something NASA should do, but rather as more of a statement of approval as much as it was a recommendation already being implemented. O'Keefe's 24 June 2004 Transformation announcement, coupled with the way Steidle had been running things simply expanded further on some examples Steidle had already set. Indeed the Commission's statement observing, "The successful development of identified enabling technologies will be critical to attainment of exploration objectives within reasonable schedules and affordable costs," is precisely what Steidle had been focusing on.

In recommending new ways of thinking and engaging the private sector in what NASA does, the Commission recommended, "NASA aggressively use its contractual authority to reach broadly into the commercial and nonprofit communities to bring the best ideas, technologies, and management tools into the accomplishment of exploration goals," and that "Congress increase the potential for commercial opportunities related to the national space exploration vision by providing incentives for entrepreneurial investment in space, by creating significant monetary prizes for the accomplishment of space missions and/or technology developments and by assuring appropriate property rights for those who seek to develop space resources and infrastructure."

By the time the Commission's report was released, NASA had already implemented its Centennial Challenges program of prizes. Indeed, a standing-room only meeting was held in Washington at the very same day and time as the Aldridge commission was releasing its final report (the overlap due to a delay in the report's release due to the death of Ronald Reagan). In addition, Steidle's Office of Exploration Systems had been using an increasing variety of procurement mechanisms—such as Broad Agency Announcements (BAA) which were not common to NASA—in an attempt not only to focus the process of seeking ideas and formulating contract solicitations, but also to widen the scope of participants and modes of participation available between NASA and the external world.

The Road Ahead

George Bush announced his new space policy in an election year. The announcement was immediately controversial to some. It was uplifting to others. As the Washington saying goes, "Where you stand depends on where you sit."

Many people are against the proposal simply because they do not like George Bush and they proceed to espouse views from that core political stance. All of the Democratic presidential candidates spoke critically of Bush's policy before and after it was announced. Those candidates who support space—but not Bush—found a way to do both.

When it comes to Congress, where this new space vision could, if enacted into reality, result in additional funds to a number of states, support for the plan tends to cross political distinctions; but does so along long established political fault lines that connect the so-called 'space states' of Texas, Alabama, Florida, Mississippi, California, Virginia, and Maryland.

Then, of course, some people are simply against the whole idea of space exploration in general and often reiterate the common refrain, "Why spend money in space when it could be better spent on Earth." Of course, none of these folks stop to ponder the notion that, as of the year 2004, all money spent on space IS indeed spent on Earth and the vast majority of the funds go to create jobs in the public and private sector. The money that goes in the pockets of these workers is just as green as that coming from other sources. Until such time as there are banks and commercial institutions located off of Earth's surface, all of this money will continue to be spent on Earth.

The 2004 Election

Usually it is rare that you hear candidates for national office make any mention of space exploration. However, with the rather substantial attention given to President Bush's speech on his new space policy before and after the announcement, the topic became a momentary lightning rod for others to use to make political jabs at Bush.

On 6 November 2003, former Governor Howard Dean (D-VT), who at the time was still the candidate to beat in Iowa, participated in an online chat sponsored by the Washington Post and the Concord Monitor. Dean was asked a question about space: "Dallas, Tex.: If elected president, what are your plans for NASA and the Space Program? Do you think it's time to retire the Shuttle and move on to bigger and better things, such as a human mission to Mars, or returning to the Moon?" Howard Dean: "I am a strong supporter of NASA and every government program that furthers scientific research. I don't think we should close the shuttle program but I do believe that we should aggressively begin a program to have manned flights to Mars. This of course assumes that we can change presidents so we can have a balanced budget again."

The balanced budget caveat would be a constant refrain by Dean in the coming weeks when the space policy issue was raised. On 11 January 2004, a few days before Bush's speech, the Marin Independent Journal wrote, "But politics does not stop at the atmosphere's edge. At a rally in Rochester, N.H., yesterday for his campaign for the Democratic presidential nomination, Howard Dean, the former Vermont governor, suggested that Bush's motivation was political, and he asked how the program would be paid for. 'I happen to think space exploration is terrific,' Dean said. 'Where is the tax increase to pay for it? It is not worth bankrupting the country.'"

On 12 January 2004 Associated Press asked Dick Gephardt (D-MO) about the impending announcement of a new space policy by the Bush Administration: "Asked about Bush's expected call for a costly new effort to return U.S. astronauts to the Moon and send them to Mars, Gephardt said that NASA ought to remain focused on the space station. 'I think we ought to see this through before we go on to something else,' he said. He also said that with the deficit at $450 billion, the nation's first priority ought to be job creation." On January 15th, the day after Bush's speech, the Washington Post wrote about Gephardt again saying, " The passion that unions feel about defeating Bush became evident when, in his speech, Gephardt made a dismissive reference to the president's new proposal that would send Americans back to the Moon and eventually to Mars. Spontaneously, the union workers broke out in a chant: 'Send Bush to Mars! Send Bush to Mars!'"

Criticism of a new Bush space policy was not limited to the more liberal portions of the Democratic party. Sen. Joe Leiberman (D-CT) also a presidential candidate at that time was asked a question about space during an appearance on CNN's Late Edition hosted by Wolf Blitzer:

BLITZER: "Let's go through a few of the substantive issues on the agenda right now. The president expecting next week, in this coming week, to announce a major new initiative on space, to perhaps send man, maybe women, men and women, back to the Moon for some sort of a permanent base there, and maybe even, long term, go out to Mars. Is this money well spent?"

LIEBERMAN: "You know, I have very mixed feelings about it, but I'll make clear where I end up. Remember, I was attracted into politics by President Kennedy, so the Moonshot program thrilled me, and I've always supported the space program. But if you ask me whether the best use of $1 trillion of American taxpayer money in the coming years is to land a mission on Mars or the Moon, I'd say no. We need it right here on Earth to give health care that's affordable to everybody, to improve our education system, and do better on veterans' benefits and homeland security. And I'll tell you, I've got an idea to create an American center for cures, that will set as the goal something that seems as impossible today as it did when Kennedy said we could go to the Moon, and that's to cure chronic diseases like Alzheimer's and forms of cancer and diabetes, et cetera, et cetera. But if we need—if we had that kind of money, we could do it right here on Earth. And, frankly, I think that's more important to the American people than that kind of space voyage at this point in our history."

Leiberman's former running mate, former Vice President Al Gore, who has had a long-standing interest in technology and space was quoted on January 15, 2004 as saying, "Instead of spending enormous sums of money on an unimaginative and retread effort to make a tiny portion of the Moon habitable for a handful of people, we should focus instead on a massive effort to ensure that the Earth is habitable for future generations."

It would seem that George Bush had no fan base amongst the Democrats.

John Kerry on Space 2004

Of course, the only comments from a Democratic presidential candidate in 2004 that have come to have any real relevance to the future progress of Bush's new space policy (should Bush lose) are those of John Kerry, the Democratic Party's 2004 nominee. The day after Bush's speech, the San Francisco Chronicle quoted Kerry as saying, "Rather than sending Americans to Mars or the Moon right now, these people would be better off trying to figure out how to get Americans back from Iraq."

On 26 February 2004 the Cleveland Plain Dealer, reporting on Senator and former Astronaut John Glenn's endorsement of John Kerry's candidacy for president wrote: "In the afterglow of pioneering astronaut John Glenn's endorsement, Sen. John Kerry said Wednesday that he would fight to create new high-paying jobs in America with the same vigor that President Kennedy demonstrated in launching the space program in the 1960s … But Kerry said the U.S. government should not be talking about returning to the Moon or going to Mars—missions proposed by President Bush. Rather, he said, leaving his prepared speech, 'We need to go to the Moon right here on Earth' by creating high-paying jobs of the future and making sure that 'young Americans in uniform are never held hostage' to Middle East oil.'"

In a written response to questions submitted by Space.com in June 2004 to Sen. Kerry's campaign, Kerry is quoted as saying, "NASA is an invaluable asset to the American people and must receive adequate resources to continue its important mission of exploration... However, there is little to be gained from a 'Bush space initiative' that throws out lofty goals, but fails to support those goals with realistic funding."

Space.com goes on to quote Kerry as saying, "The civil space program acts as an engine of innovation for the entire country, making its enormous benefits hard to quantify but even harder to discount... I'm excited by potential advances in pharmaceuticals that microgravity could lead to… Unique drug treatments produced in the microgravity environment may play a vital role in reducing the cost of health care and in developing defenses against chemical and biological terrorist attacks."

Kerry has gone on the record favoring, at least as a general notion, the possibility of raising NASA's budget. On the official Kerry campaign website, a 28 August 2003 article titled 'John Kerry's Plan to Fight for America's Economic Future' says: "Kerry will fight to connect every American family to the Internet, encourage a renewed educational focus on science and math, bring the best practices of operational efficiency from the private sector to the public sector, and restore the government's commitment to scientific achievement through increases in research funding for the Department of Energy, NASA, and the National Science Foundation."

A trillion here, a trillion there ….

However, with regard to the Bush space policy, it would seem that the Kerry campaign was a little too eager to try and cast doubt on budget numbers used by the Bush Administration—and thus undermine the credibility of whatever Bush proposed to do in space.

In a 5 April 2004 official press release titled 'New Report Reveals $6 Trillion in

Hidden Spending in Bush Budget', the Kerry campaign says, "The True Cost of the Mars Mission ($160 billion to $1 trillion): President Bush has only included $1 billion in increased NASA funding to fulfill his ambitious plan to establish a lunar base and land people on Mars. Independent estimates of the cost of the Mars mission range from $160 billion to $1 trillion.[3]" The $1 trillion reference is listed as "[3] The $160 billion estimate is from Congressional Testimony by Michael Griffin, former Chief Engineer of NASA on 3/10/04. The $1 trillion estimate is from Gregg Easterbrook, 'Red Scare,' The New Republic, 2/2/04."

In Gregg Easterbrook's 2 February 2004 article 'Red Scare,' cited as a source for a mission cost estimate by the Kerry campaign, you can see that Easterbrook uses only conjectural semantics—and not any actual costs, specific designs, or mission architectures, to imagine that the cost of a single mission to Mars would be $1 trillion. How Easterbrook arrived at the number is simply not revealed. By the tone of Easterbrook's article, it is clear that he does not like the Bush space policy. As such, he made up some sticker-shock to help him make his case.

The interesting thing about this $1 trillion figure is that you can never trace it to a factual cost analysis. Instead it is a large round number that is used to scare people— or when reporters haven't done their homework. It goes around like a virus too. In January 2004, former Associated Press reporter Paul Recer included it in an article he wrote, but could never identify its source.

Former Columbia Accident Investigation Board panel member Douglas Osheroff made a similar unsubstantiated comment in another Associated Press article by reporter Ted Bridis on 14 June 2004 where he is quoted saying, "Never let it be said that NASA tends to overestimate the cost of its missions," said Douglas Osheroff, a widely renowned physicist who investigated the Columbia accident. "The cost in present-day dollars ... I think it's going to be one trillion." Osheroff 'thinks' it will cost a trillion dollars. Bridis also makes reference to the $1 trillion figure—but never apparently sought to confirm it for himself or ask Osheroff what his source was. A few days after this article appeared Osheroff was one of 48 Nobel laureates who formally endorsed John Kerry.

The net result of this urban myth is that everyone seems to think that a cost estimation was made—by someone, somewhere—that said the new Bush space policy was going to cost $1 trillion when in fact no such estimate has ever actually been derived.

Kerry's past record on space

Space, as a campaign issue, has been, at best, a blip on the political radar screens of national campaigns. Whether it will become an issue in the 2004 race is unclear, but doubtful. While Kerry has spoken little about space during the 2004 race, he has taken a very public stand during his career in the Senate. It is from his previous statements that one can get an idea of where he once stood, and likely still stands vis-à-vis space. Should Kerry win the election, it would be logical to expect that his stance on space as president would at least use previous statements as a departure point.

Perhaps the clearest insight into John Kerry's view on human spaceflight can be seen in his floor statements from the Congressional Record. One 1996 floor statement in regard to HR 3666 and an attempt to cancel funding for the International Space Station is particularly illustrative. Curiously, while Kerry expressed excitement for the prospect that drugs could be created in microgravity (one of the things ISS was supposed to do) in June 2004, a decade earlier, Kerry justified canceling the ISS in favor of a stronger focus on fighting disease:

SPACE STATION FUNDING (Senate - September 04, 1996)

Mr. KERRY. Mr. President, I join with the distinguished Senator from Arkansas as a cosponsor of his amendment and urge my colleagues to support this effort to terminate funding for the National Aeronautics and Space Administration Space Station program, which the General Accounting Office estimates will cost American taxpayers $94 billion.

Every day, the working families of Massachusetts have to make tough choices about what they can afford, how to pay the rent, and whether they can send their kids to college.

The Federal budget deficit, while reduced by two-thirds due to President Clinton's leadership and the courage of the Democratic-controlled Congress in 1993, is still too high and must be eliminated. It is a drain on our economy and, increasingly, the debt service we pay is robbing us of the ability to make badly needed investments in our future. I have been working in the U.S. Senate to make the tough choices necessary to balance the budget.

When measured against this imperative, I believe the space station's potential benefits—which I recognize—do not stand the test. I believe we must terminate funding for this program.

We cannot spend nearly $100 billion of the taxpayers money to fund the space station and then say that we do not have enough money to put cops on the beat, clean our environment, and ensure that our children get the best education possible.

The Senator from Arkansas, joined by several others of us, has made a valiant effort to halt this project again and again over the past several years. I am hopeful that this year the time has come when the Senate will exercise fiscal responsibility over our Federal budget, like any family in Massachusetts would over its own family budget, by terminating the space station immediately in order to reduce the deficit.

In 1984, NASA justified the space station based on eight potential uses. Now only one of these assignments remains: the space station will be used as a research laboratory. However, the costs of performing scientific research in space simply outweigh the potential benefits. It will cost over $12,000 to ship 1 pound of payload to the space station.

Many of my colleagues support the space station because it creates jobs. But the project's costs for developing jobs are exorbitant—those jobs will cost approximately $161,000 each. If invested here on terra firma, that amount of money would fund three or four or even more jobs.

As a member of the Senate Commerce Committee, I have fought, along with the distinguished Senator from South Carolina [Mr. Hollings] and other Senators, to secure funding for many important scientific programs. Many of these programs have been shortchanged in order to help pay for the costs associated with the development of the space station. Allowing this extraordinary large science program to receive funding at the expense of these other so-called small science programs—which I believe will produce more products and more valuable products—is unacceptable. These small programs are creating thousands of high wage technology jobs at a fraction of the cost associated with the space station.

In the space program itself, the enormous level of funding consumed by the space station is crowding out much smaller programs for satellites and unmanned space probes, which most experts consider more cost-effective than manned missions.

These activities are aimed at expanding our understanding of the Sun, the solar system, and the universe beyond. The specific programs in this category include the 'new millennium,' a program to build robotic spacecraft one-tenth the size and cost of satellites; the Cassini mission to Saturn, scheduled for launch in 1997; continuation of the Discovery missions, each of which costs less than $150 million, can be launched within 3 years of the start of its development, and is used by NASA to find ways to develop smaller, cheaper, faster, better planetary spacecraft; and the Mars surveyor program which funds a series of small missions to resume the detailed exploration of Mars after the loss of the Mars Observer mission in 1993.

Funding for projects in this area will be approximately $1.86 billion in fiscal year 1997 which represents a 9-percent reduction from last year. The academic research establishment is concerned that the space station appears to be draining funds from these other space projects.

Also included among the programs placed at risk by the space station is the mission to planet Earth, NASA's satellite program to explore global climate change by means of a series of Earth observing satellites launched over a 15-year period, beginning in 1998—a program endorsed by the National Academy of Sciences.

Given the structure of congressional appropriations bills, the enormous funding for the space station has come not just at the expense of other space programs but at the expense of environmental research and other important activities that promise to improve the lives of our citizens and enhance our security more completely.

Building the space station has become a joint effort between the United States and Russia. We all want to see continued progress in United States-Russian relations. However, we should be encouraging Russia to house and feed its own people, provide jobs, and above all care for its deteriorating nuclear power plants and dismantle its nuclear missiles and warheads. Asking Russia to commit its resources to pursue an uncertain and risky space station venture instead of encouraging it to tend to these important matters is unwise.

Some may argue that we have lost our vision if we terminate the space station. But their concern is misplaced. We still have vision. But the vision is to restore the American dream to our citizens, to restore their sense of safety on the streets, to invest in technology that will increase our competitiveness and the quality of jobs, to invest in research that will cure our deadly diseases, and to restore our communities to the condition where children can learn and dream.

It is time to decide. I think the American people are watching impatiently to see whether the U.S. Congress can deliver spending reductions for programs that are politically popular but fiscally unwise.

I commend my distinguished colleague from Arkansas, Senator Bumpers, for his continuing leadership on this important issue. I urge all my colleagues to vote to terminate the space station."

Kerry expressed similar sentiments in regard to the FY 1994 VA HUD Appropriations Act on September 21, 1994:

"We have libraries and schools in the United States of America that are shut in the afternoon and kids have nowhere to go. We have whole cities that are deprived of Boys and Girls Clubs so only 10 percent of the population has a place to find an outlet. But we can find money to put a few astronauts up in space at this moment in time?

I would love to do that. I was raised on the promise of President Kennedy. Someone here asked earlier, 'Don't we have people of vision anymore?' Yes, Mr. President, we do. But the vision is to restore the American dream to our citizens, to restore their sense of safety on the streets, to invest in technology that will increase our competitiveness and the quality of jobs, to invest in the research that will cure our deadly diseases, and to restore our communities to the condition where children can learn and dream.

Will terminating this program hurt in California? Will it hurt in Texas? Will the loss of $600,000 hurt in Massachusetts? Yes, it will hurt. But if we measure that loss against the pain that people across the country are feeling because we are not willing to address our fundamental needs as a Nation.

It is hard choice time. That is what this is about, and I think the American people are waiting feverishly to see whether the United States Congress can actually do something for once—whether we can really deliver some spending reductions and make some of the choices we ought to make for our future.

Mr. President, I hope we will finally ante up and deliver to the American people. I had a separate bill to try to cut the space station and a number of other wasteful Federal programs. I am delighted to join the Senator from Arkansas and the Senator from Tennessee and others who are leading in this effort to try to help the Congress do the responsible thing. I hope we will succeed."

John Kerry has a generally supportive view of NASA—as long as he doesn't have to get nailed down to specifics. This is not to say that he would not support the agency overall—indeed he has indicated that he would. But when it comes to the International Space Station, his voting record speaks resoundingly clearly:

In 1991, Kerry Voted To "Reduce Funding For The Space Station From $2 Billion To $100 Million," And Transfer Funds To Other Programs. (H.R. 2519, Congressional Quarterly Vote #132: Rejected 35-64: R 3-40; D 32-24, July 17, 1991, Kerry Voted Yea)

In 1992, Kerry Voted To Terminate Space Station "Freedom" Project. (H.R. 5679, Congressional Quarterly Vote #194: Rejected 34-63: R 4-39; D 30-24, September 9, 1992, Kerry Voted Yea)

In 1993, Kerry Voted "To Terminate The Space Station Program." (H.R. 2491, Congressional Quarterly Vote #272: Motion Agreed To 59-40: R 36-8; D 23-32, September 21, 1993, Kerry Voted Nay)

In 1993, Kerry Voted To Terminate Space Station Program And Divert Funds To Tax Cuts. (H.R. 3167, Congressional Quarterly Vote #335: Motion rejected 36-61: R 10-32; D 26-29, October 27, 1993, Kerry Voted Yea)

In 1994, Kerry Voted To Cut $1.9 Billion From Space Station Program, Thus Terminating It. (H.R. 4624, Congressional Quarterly Vote #253: Rejected 36-64: R 6-38; D 30-26, August 3, 1994, Kerry Voted Yea)

In 1995, Kerry Voted To Reduce NASA Funding By $400 Million. (H.R. 889, Congressional Quarterly Vote #105: Motion Agreed To 64-35: R 43-11; D 21-24, March 16, 1995, Kerry Voted Nay)

In 1995, Kerry Voted To Cut $1.8 Billion From NASA's Human Space Flight Program. (H.R. 2099, Congressional Quarterly Vote #463: Motion Rejected 35-64: R 12-41; D 23-23, September 26, 1995, Kerry Voted Yea)

In 1996, Kerry Voted To Cut $1.6 Billion From NASA's Human Space Flight Program And Terminate Space Station Program. (H.R. 3666, Congressional Quarterly Vote #267: Motion Agreed To 61-36: R 38-12; D 23-24, September 4, 1996, Kerry Voted Nay)

Given this rather blunt rejection of human space flight and a permanent human presence in space, one has to wonder: if Kerry is this strongly against the International Space Station, a multi-year, multi-billion dollar international program several hundred miles overhead, whether he'd be any more interested in a similarly large, long-term project that sent humans to the Moon or Mars.

Then again, people have been known to change. Nixon went to China.

Will George Bush support his own policy?

As for the prospects that George Bush will support George Bush's space policy, this of course, depends on whether he is re-elected. If he is, then the classic forces that shape second presidential terms come into play: without fear of re-election, the president can afford to make excursions that might be more politically risky. He might move into legacy mode and be thinking of things that he wants to leave behind— things that may not have a lot of political gravitas at the time they need to be put forth.

George Bush's space policy was announced in an election year and ran the risk of being politicized and/or ignored, thus allowing it to lose steam. But the policy also had the benefit of a running start, of sorts, that Sean O'Keefe had been putting in place in the two years prior to its announcement. To be certain the Columbia accident gave NASA's credibility a body blow but it also allowed many to muster emotions and a commitment that might not have been possible had there not been such a tragedy to focus everyone's attention.

The technical challenge of turning George Bush's vision into the reality of hardware was well understood. It was also a bit daunting. NASA and its contractors had not done anything like this in a generation, but they had gone this way before.

But the more difficult task would be not technical but rather, political. George W. Bush had chosen to govern in a far different way than his father, different even than the president to which he had often compared himself, Ronald Reagan. Both Reagan and George H. W. Bush had sought consensus in building their legislative and programmatic coalitions. Reagan, for all of his pre-White House reputation as a hard-edged conservative governed very differently. While never compromising his core principles, Reagan had often sought compromises as he built his tax and budget programs, understanding that change could also be achieved in incremental steps. The 'Great Communicator' also knew how to use his public persona to advance his agendas, cloaked in less threatening language and deconstructed for public consumption.

George W. Bush had promised to govern as a 'uniter, not a divider'. But by comparison to Reagan, he had done just the opposite in many instances. Perhaps it was decided that sharp-edged partisanship was the formula to win the mid-term election of 2002. Perhaps it was the legacy of the 2000 Florida recount. But whatever the reason, his approach to passing his policies was more often based on obtaining a majority from the Republican caucus first, then, and only then if needed, reaching across the aisle.

His science policies, from global warming to stem cell research, were the focal point not of broad-based acceptance by the science community, but accusations of bias and a pre-existing slant towards conservative views of each issue. His science advisor, Dr. John Marburger, a widely respected Democrat, fought the accusations of political taint to Bush's programs. But it would linger throughout the period when Bush would normally have needed support from some of these groups for his expanded space agenda.

What Bush found instead was either indifference or opposition from some who might under other circumstances support the idea of advanced space goals. Former Ohio Senator and Mercury/Shuttle astronaut John Glenn would become a voracious critic of the Bush space plan. Glenn used the cuts to space station research as the public basis of his opposition. But beneath the surface, politics lingered, as Glenn endorsed Senator John Kerry's bid to replace Bush, and Kerry was no friend of the Bush space plan—or of human spaceflight.

Donna Shirley was another person whose support would, based on her NASA and science career, be expected. But Shirley, a Democrat and supporter of Democratic causes, had refused to sign on to endorse the Bush program. And there would be others, who might well be enthusiastic about a return of America to space faring, if the name attached to the plan was someone other than George W. Bush.

The price of polarization would stalk the Bush space policy as it made the rounds of Capitol Hill and the country. O'Keefe's task was to get Congress, and the public at large, to look afresh at the plan and how it would evolve and be funded, not focus on its political parentage. In the end, though, any presidential policy cannot be separated from its origins, for rhetoric cannot trump politics.

After O'Keefe – who?

Of course, there is the question as to who would sit at the helm of NASA – even if Bush wins a second term. O'Keefe, a loyal republican, had accepted the request to serve as NASA Administrator on the spot in the president's office without taking a moment to ponder it. It is tradition that all White House employees formally tender their resignation at the end of an administration. Once in a while, the letters are accepted. Other times, people are moved around. O'Keefe is habitual in his carefully worded, succinct reply when asked how long he'd serve at NASA saying that his appointment letter says that he "serves at the pleasure of the president - for the time being."

O'Keefe has been moved from one job to another in several Administrations. Few would be surprised if it happened again. Nor would people who know him be surprised if he stayed put because the president wanted him to stay.

It is no secret within the agency that people genuinely like O'Keefe and that his personal charm is often the tool he uses to convince people to do things a certain way. That said, O'Keefe's ongoing challenge has been to try and create a management ethos and structure that is independent of personality. A logical point at which you might see a change in NASA leadership (during a Bush second term) would be at the point that the space shuttle fleet resumes operations. While it is obviously premature to speculate who Bush might fill O'Keefe's office with—or even if he'd even consider doing so—it is instructive to look at how Bush did it last time.

In 2001, the agency was in a financial and managerial mess. So the White House put one of their top financial and managerial people over at NASA to fix it. Now the agency is back on its feet and is healthy enough such that the White House feels it can be given a new challenge. As such, it would be reasonable to expect that they would look for someone with the technical and programmatic leadership qualities needed to take a repaired agency into the 21st century.

While no indication whatsoever has been heard to suggest that a replacement for O'Keefe is contemplated by Bush, or that names have been circulated, an obvious type of person that might be considered would be Craig Steidle, the man O'Keefe brought to NASA to manage the implementation of the president's new space vision.

As for who John Kerry might put in the job; based on Kerry's track record and recent statements, it would have to be someone who did not have a problem presiding over an agency whose people now needed to be told that the plans they had gotten so excited about were now under re-evaluation.

An uncertain road ahead

As 2004 headed into mid-year, the fate of the space plan was still uncertain. But the process—the 'journey'—that had led to the policy was among the most comprehensive and rigorous any space proposal had ever endured. Many previous space policy proposals; from Kennedy's lunar commitment of 1961, to Nixon's reluctant endorsement of the space shuttle in 1972, to Reagan's permanent space station decision in 1984 had run a similar gauntlet.

But none had yielded the multi-dimensional program that Bush's exercise had produced. And none previously had attempted to justify their plans by, in essence, cannibalizing the very bureaucracy that would lead the programs.

It was a revolution, a transformation of an agency that had fought change at the core of its being. It took a national catastrophe that would disgrace NASA, a president unwilling to ignore the ramifications of the catastrophe, and a political insider to cobble together the thread of revolution in space during a time of war and deficits. Sean O'Keefe, Paul Pastorek, Craig Steidle, Steve Isakowitz, Ed Weiler, and Bill Readdy all knew the implications of what Bush proposed. But a revolution cannot be sustained by only one set of revolutionaries. For organizations to be truly changed, transformed, the organization must want this change at it innermost level—the very people who make the place run.

O'Keefe's reformers knew this, of course. They had rode the shifting political winds to power and had made the most of the climate they had faced. It remained to be seen if those same political winds would shift again, in a process of renewal and reinvention that were hallmarks of American government. If so, then much of the struggles of 2003 and 2004 to launch the new space vision would be stillborn.

That would produce, not the massive new space exploration agenda, but what Paul Pastorek called 'a Camelot moment', a sweet memory of an attempt to do good government that failed because of the political price its parent paid at the voting booth. If that would be the legacy of the space vision exercise, then it was not clear that any future American president would risk his or her political capital or prestige to travel that way again—at least not for another generation.

A moment in time

Sean O'Keefe often spent the night at Paul Pastorek's apartment near NASA's E Street headquarters, if the weather or the lateness of the hour made the trip home to the Virginia suburbs unwise. On one such night in August 2003, O'Keefe and Pastorek stood on the balcony of the apartment and shared a scotch or two. The hours passed, and the summer night sky was clear and filled with stars. The two old friends were quiet, thinking of the historic events that were unfolding around them, for the CAIB had just released its report.

Then, after a time, O'Keefe turned to Pastorek. "Here we are, dealing with a national crisis," he said to his old college friend as the star-filled night sky cartwheeled above them. "A couple of baloneys from Loyola."

Change agents. Or not. Stay tuned.

*　　　　　　*　　　　　　*　　　　　　*　　　　　　*

EPILOGUE:

LEGEND AND LEGACY

Some two thousand miles to the north of where Scott Thurston stood watch over Columbia's remains, the first anniversary of its accident approached. It was clear that its observance would require reliving the anguish of the event. In a sense, that anguish had never really gone away. During every meeting, discussion, hearing— wherever space people gathered—the wound of the accident always lay in the background, unsaid, unacknowledged, but still there just the same. Long after the flags returned to full staff, the wreckage retrieval had ended, and the CAIB report had been released— NASA was en route to 'Return To Flight'.

Yet there was always the inevitable memory of the lost shuttle—and its crew— lingering in the back of one's mind.

After the accident, spontaneous memorials popped up all across NASA, the U.S., and the world. Most of these faded as the flowers wilted and the rain and the wind had washed them away. Some small towns—most notably a number in Texas—erected more permanent memorials. To their citizens and the people of west Texas, the space program and the shuttles had become almost a part of their family. Others marked the passing of the craft and its crew in different ways.

The summer after the accident, it was announced that a peak in the Sangre de Cristo Mountains in Colorado was to be named 'Columbia Point' in honor of the crew. The location of this peak is on the east side of Kit Carson Mountain. On the northwest shoulder of Kit Carson Mountain is 'Challenger Point', a peak previously named in memory of the first lost shuttle.

A few weeks later, on Devon Island, in the Canadian arctic, members of the NASA Haughton Mars Project erected a series of memorials—one for each member of the crew. The structures took the form of an 'inukshuk'—large, free-stone structures built in the rough shape of a human form. The local people, the Inuit, had been erecting inukshuks as navigation markers for centuries. Once built, they could last for many generations.

In January 2004, NASA announced that a plaque had been placed on both the 'Spirit' and 'Opportunity' Mars Rovers. With the successful landing of the Spirit on Mars, the words on the plaque were revealed: "In memoriam to the crew of the Space Shuttle Columbia." All of the crew's names were listed along with the STS-107 mission patch, the NASA logo, and an American flag. An Israeli flag appeared next to the name of Ilan Ramon. The location where Spirit landed was hereafter to be called 'The Columbia Memorial Station'.

After the crew of Columbia had been laid to rest and commemorated in many ways, a single, national memorial was created for dedication as the accident's first anniversary drew near. On the morning of February 2, 2003, a memorial ceremony was held at Arlington National Cemetery. Near the place which Sean O'Keefe, Bill Readdy, Fred Gregory, and Bryan O'Connor had visited, almost a year before, hundreds now met to commemorate a stone and bronze memorial to mark the Columbia's loss.

Sean O'Keefe sought to place this memorial—next to one honoring the crew of the shuttle Challenger—into context. Speaking into the cold winter wind, he said, "Arlington National Cemetery also provides a final resting place for great heroes who changed the course of history by blazing new trails of exploration and discovery. Among those honored at Arlington are such legends of exploration as John Wesley Powell, the first man to explore the Grand Canyon, Admiral Richard Byrd, the first to fly over both poles, and the discoverers of the North Pole, Robert E. Peary and Matthew Hensen. Resting here at Arlington is also the president who boldly set our course to the stars, John F. Kennedy."

O'Keefe's remarks were brief, intense, and below the surface emotional. At the end, he pointed toward the future. "Generations from now, when the reach of human civilization is extended throughout the solar system, people will still come to this place to learn about and pay their respects to our heroic Columbia astronauts. They will look at the astronauts' memorial and then they will turn their gaze to the skies, their hearts filled with gratitude for these seven brave explorers who helped blaze our trail to the stars."

Before and after the ceremony hymns were played by several military bands. The faces in the crowd were somber, many wept openly.

Later that afternoon, NASA would announce that a series of hills on Mars—visible on the horizon to the west of where Spirit had landed—would hereafter be known as the 'Columbia Hills'; with each promontory named after a member of its last crew. Spirit would later begin a month-long trek toward these hills once it had completed its primary mission. Later, three hills in this area would be named in honor of the Apollo 1 crew. A month or so later, once the Rover 'Opportunity' had landed on Mars, its landing site would be named 'The Challenger Memorial Station'.

The ceremonies and dedications to the crew of Columbia were painful to the space community. But they were set along familiar themes that Americans have come to know as traditional. But on the day following the dedication, a memorial was held in Washington, DC which marked a wholly different national tradition.

Several hundred invited guests gathered at the Embassy of Israel on that cold, wet night to remember Ilan Ramon. Daniel Ayalon , Israel's Ambassador to the United States began the event by recalling his pride at the launch of the mission. He talked of Ilan as the son of a holocaust survivor, a veteran of many dangerous missions in the defense of the Israeli nation, and the country's first astronaut. His story, he said, epitomized the story of Israel and the Jewish people. The entire country had been waiting for Columbia to return, and Ayalon said, the pain of its loss would always be with them.

Elizer Zanberg. Minister of Science and Technology for the state of Israel spoke, and was followed by cellist Amit Peled, who played Saraband from the suite for cello solo in C major, by JS Bach.

In marked contrast to the traditional military hymns played at Arlington the previous day, Peled's performance pulled at the soul. One could imagine all of the people in the room being drawn into every note; as if being spoken to by the cello. Then it was time for Sean O'Keefe to speak, on behalf of the president, the nation, and NASA.

He talked of heroes.

"We at NASA are extraordinarily privileged to have heroes the caliber of Ilan representing all humanity as we extend our grasp into the vast reaches of God's creation," O'Keefe said. He spoke of Ramon's passion for mountain climbing, of the new Columbia Point, of his children's determination to climb it. Last summer they, along with the family members of the other Columbia astronauts, participated in an attempt to climb that new landmark. But a storm prevented them from getting to the 14,000-foot summit, O'Keefe said. But while on the mountain they were able to hold a remembrance ceremony at 13,100 feet, the same height that the crew ascended to on Wind River Peak in Wyoming during their training. High above them a special jet flyby swept across the mountain range, a tribute to the crew. He then saluted all of Ramon's children for helping to carry on the legacy of their father's adventurous spirit.

At the end of the event, Rona Ramon, Ilan's widow, spoke last. Steeling her emotions with grace and clarity, she spoke elegantly and briefly. She thanked all for coming. And then she talked of her husband, and the flight of the lost shuttle. "Our mission in space is not over," she told the hushed audience. "He was the first Israeli in space— that means there will be more."

Out at Arlington, the bas-relief of the shuttle memorials would be seen by thousands of visitors who braved the snow and ice that winter. Visitors to the cemetery now could pause between space memorials, both built in sorrow to mark the loss of two machines and their crews. Challenger and Columbia had been destroyed by mechanical failures, to be sure.

But what lay beneath those malfunctions was hubris, a failure to understand and to listen. A need to return to such basics were at the heart of the reforms O'Keefe was attempting to impose on his reluctant agency. A change, he thought, to earn the right to execute President Bush's vision, which at it's center was a simple but radical idea—return the space program and the space agency back to its roots of exploration, discovery and risk-taking. His team would seek to make it a less insular, closed place. Listen, and open the doors to fresh ideas. And, using those ideas, take the next generation to what Neil Armstrong would call "places to go beyond belief."

Bush saw it as exploring. O'Keefe saw it as a chance to transform his agency. Perhaps, in the end, its last chance to be transformed. On February 1, 2003 a hinge of history opened, and with it the possibility for new achievements in space. That possibility had been purchased for the nation in blood and treasure. O'Keefe was determined, as was his team of would-be reformers, to make certain that its cost, and who paid it, were never forgotten.

On one cold dark Sunday afternoon not long after the dedication, a mother and her young son came to the astronaut's graves. Her identity is unknown to the authors of this book—but her actions were witnessed by one of us. Speaking with a strong accent, she tried to describe to her child what these plaques meant, and who lay buried nearby. "They were heroes," she tried to explain, who had voyaged into space aboard the space shuttles "to help the world learn new things about the universe," about space, but had died in the effort.

But the little child didn't understand what she was trying to convey. "Why, mom?" he asked repeatedly, "Why did they die?" She struggled again to make sense of it so he could grasp what it was they were seeing, in a way that would fix the moment in his memory. But the words didn't work. In the soft, gathering winter gloom the child just didn't understand. Finally, after he asked the second or third time, "Why did they die, mom?" his mother gave up trying to explain it.

But after a moment she looked down at her son. As the sky high above Arlington turned into sunset, she said softly, "For you."

*   *   -   *   *

ACKNOWLEDGEMENTS

The authors wish to acknowledge the help and encouragement of many people during the course of researching and writing this book. Many who played critical roles in making themselves available for interviews wished, for their own reasons, not to be identified. We have therefore omitted their names but have not omitted the debt that we have incurred for their help and wise counsel. Without them, there would, literally, be no book.

We gratefully acknowledge the help of Sean O'Keefe, Fred Gregory, Paul Pastorek, Bill Readdy, Craig Steidle, Michael Hawes, Bob Jacobs, Robert 'Doc' Mirelson, Allard Beutel, Brendan Curry, Al Feinberg, Laura Brown, Jessica Rye, Scott Thurston and Glenn Mahone in providing access and support. While the senior leadership of NASA provided us much needed access, they in no way have endorsed this book, or had any opportunity to review its contents prior to publication. The sole exception was in reviewing direct quotes of some of the participants, to assure accuracy. We also gratefully acknowledge the help of Frank Morring in the preparation of the final manuscript. Frank, we owe you much more than a thank-you. NASA headquarters audio-visual office also helped gather together the videos that are contained in the DVD in the rear pocket. Without them, there would be no DVD.

For moral support and many good ideas, Frank and Keith wishes to thank Phil Berardeli, Elaine Cami, Scottie Barnes, Irene Klotz, Kate Kronmiller, Don Brownlee, Lynn Heninger, Tidal W. McCoy, Rich Coleman, Suzie Chambers Sterner, Melissa Sabatine, Susan Flowers, Laura Brown, Pat and Julius Dasch, Brian Chase, Sandra Brubaker, Jim Muncy, Wayne Kubicki, Sue Hegg, Ian W. Pryke, Melinda Gipson, Charlie Walker, Patricia Grace Smith, the late F. Kenneth Schwetje, James R. Kirkpatrick, David S. Schuman, Jim Oberg, Craig Couvault, Michael Cabbage, Dennis Wingo, Shelby Spires, Chris Shank, Frank Braun, Maggie Nelson, Steve Hoeser, Courtney Stadd, Richard Buenneke, Jr., Glen Golightly, Charlie Barthold, Klaus Heiss, and Dee Divis.

We also wish to thank the Ace Transcription Service of Washington, D.C.

Keith wishes to thank his wife Jenny for putting up with his absence from normal life as this book was being written, to his business partner Marc Boucher, and to his parents Ken and Louise for doing all they could to foster his early interest in space. Keith would also like to thank the thousands of NASA Watch readers who have continued to provide information about the inner workings of the agency over the years.

We are both grateful for the help and support of our publisher, Robert Godwin, as well as Ric Connors, and the staff of Apogee Books for their patience and forbearance with writers who are incapable of meeting deadlines.

For all of the support, suggestions, and participation direct and indirect of these and many others, we will always be grateful. But any errors of fact, or of omission or commission, are the author's own.

Frank Sietzen Jr., Arlington, Virginia   Keith L. Cowing, Reston, Virginia

fall 2003-summer 2004

SOURCE NOTES

The core material on which the book is based is more than 12 hours of interviews with senior leaders of the civil space program and the White House. In addition, we have used the report of the Columbia Accident Investigation Board (CAIB) all six volumes of which were graciously made available by Laura Brown. We have also used the report of the President's Commission on the Implementation of Space Exploration Policy, and for that we also thank Susan Flowers.

We have also prepared a series of appendices. These included the text of President Bush's January 14, 2004 address on space policy, the White House summary of the new space policy, Bush's response letter to the Aldridge Commission report, as well as sources that could be identified for statements in the various chapters. Therefore, our publisher has decided to set up a web site for the book where this material will be available to readers.

http://www.apogeebooks.com/moonadds.htm

We may also publish additional updated information and other material on this site that were not included in the printed book.

-The authors

*   *   *   *   *

Check us out on the Web -
http://www.apogeebooks.com

Return this completed form and become eligible to win free books!

One new space book almost every month! The world's number one space book publisher!

NO POSTAGE NECESSARY IF MAILED IN THE UNITED STATES

BUSINESS REPLY MAIL

FIRST CLASS MAIL   PERMIT NO. 350 WHEATON IL

POSTAGE WILL BE PAID BY ADDRESSEE

COLLECTOR'S GUIDE PUBLISHING INC
P.O. BOX 4588
WHEATON, IL 60189-9937

I.II.iiII......iIII.iiII.iII.iIII.iIII.......IIIi.iiIIiiII

Check us out on the Web -
http://www.apogeebooks.com

Return this completed form and become eligible to win free books!

One new space book almost every month! The world's number one space book publisher!

NO POSTAGE NECESSARY IF MAILED IN THE UNITED STATES

BUSINESS REPLY MAIL

FIRST CLASS MAIL   PERMIT NO. 350 WHEATON IL

POSTAGE WILL BE PAID BY ADDRESSEE

COLLECTOR'S GUIDE PUBLISHING INC
P.O. BOX 4588
WHEATON, IL 60189-9937

I.II.iiII......iIII.iiII.iII.iIII.iIII.......IIIi.iiIIiiII

APOGEE BOOKS

THE SPACE BOOK COMPANY!

P.O. BOX 4588
WHEATON ILLINOIS 60189-9937
HTTP://WWW.APOGEEBOOKS.COM

Apogee Books is THE space book company, with almost one NEW space related book published every month.

☐ *To receive a catalog by mail check here and print your address below.*

☐ *Please update me by email (Print email address below)**

NAME: (Please PRINT)

Company:

Address:

City: **State:**   **ZIP:**

E Mail:

*CG Publishing will not disburse this email address to any other organizations for unrelated purposes except as marked below.

Are you are interested in hearing from any Space Advocacy Groups? Check the boxes below.

Space Frontier Foundation ☐   Planetary Society ☐   National Space Society ☐   All available ☐

1

APOGEE BOOKS

THE SPACE BOOK COMPANY!

P.O. BOX 4588
WHEATON ILLINOIS 60189-9937
HTTP://WWW.APOGEEBOOKS.COM

Apogee Books is THE space book company, with almost one NEW space related book published every month.

☐ *To receive a catalog by mail check here and print your address below.*

☐ *Please update me by email (Print email address below)**

NAME: (Please PRINT)

Company:

Address:

City: **State:**   **ZIP:**

E Mail:

*CG Publishing will not disburse this email address to any other organizations for unrelated purposes except as marked below.

Are you are interested in hearing from any Space Advocacy Groups? Check the boxes below.

Space Frontier Foundation ☐   Planetary Society ☐   National Space Society ☐   All available ☐

2

RENEWALS 458-4574

DATE DUE

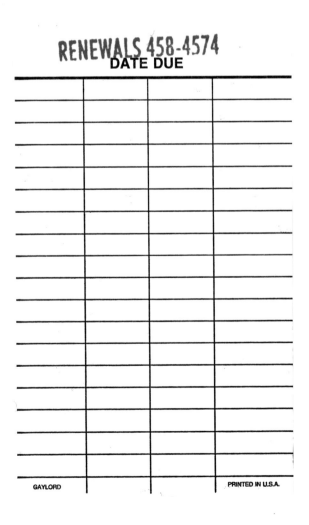

GAYLORD   PRINTED IN U.S.A.